Telescopes: A Very Short Introduction

VERY SHORT INTRODUCTIONS are for anyone wanting a stimulating and accessible way into a new subject. They are written by experts, and have been translated into more than 40 different languages.

The series began in 1995, and now covers a wide variety of topics in every discipline. The VSI library now contains over 500 volumes—a Very Short Introduction to everything from Psychology and Philosophy of Science to American History and Relativity—and continues to grow in every subject area.

Very Short Introductions available now:

ACCOUNTING Christopher Nobes
ADOLESCENCE Peter K. Smith
ADVERTISING Winston Fletcher
AFRICAN AMERICAN RELIGION Eddie S. Glaude Jr
AFRICAN HISTORY John Parker and Richard Rathbone
AFRICAN RELIGIONS Jacob K. Olupona
AGEING Nancy A. Pachana
AGNOSTICISM Robin Le Poidevin
AGRICULTURE Paul Brassley and Richard Soffe
ALEXANDER THE GREAT Hugh Bowden
ALGEBRA Peter M. Higgins
AMERICAN HISTORY Paul S. Boyer
AMERICAN IMMIGRATION David A. Gerber
AMERICAN LEGAL HISTORY G. Edward White
AMERICAN POLITICAL HISTORY Donald Critchlow
AMERICAN POLITICAL PARTIES AND ELECTIONS L. Sandy Maisel
AMERICAN POLITICS Richard M. Valelly
THE AMERICAN PRESIDENCY Charles O. Jones
THE AMERICAN REVOLUTION Robert J. Allison
AMERICAN SLAVERY Heather Andrea Williams
THE AMERICAN WEST Stephen Aron
AMERICAN WOMEN'S HISTORY Susan Ware
ANAESTHESIA Aidan O'Donnell
ANARCHISM Colin Ward
ANCIENT ASSYRIA Karen Radner
ANCIENT EGYPT Ian Shaw
ANCIENT EGYPTIAN ART AND ARCHITECTURE Christina Riggs
ANCIENT GREECE Paul Cartledge
THE ANCIENT NEAR EAST Amanda H. Podany
ANCIENT PHILOSOPHY Julia Annas
ANCIENT WARFARE Harry Sidebottom
ANGELS David Albert Jones
ANGLICANISM Mark Chapman
THE ANGLO-SAXON AGE John Blair
THE ANIMAL KINGDOM Peter Holland
ANIMAL RIGHTS David DeGrazia
THE ANTARCTIC Klaus Dodds
ANTISEMITISM Steven Beller
ANXIETY Daniel Freeman and Jason Freeman
THE APOCRYPHAL GOSPELS Paul Foster
ARCHAEOLOGY Paul Bahn
ARCHITECTURE Andrew Ballantyne
ARISTOCRACY William Doyle
ARISTOTLE Jonathan Barnes
ART HISTORY Dana Arnold
ART THEORY Cynthia Freeland
ASIAN AMERICAN HISTORY Madeline Y. Hsu

ASTROBIOLOGY David C. Catling
ASTROPHYSICS James Binney
ATHEISM Julian Baggini
AUGUSTINE Henry Chadwick
AUSTRALIA Kenneth Morgan
AUTISM Uta Frith
THE AVANT GARDE David Cottington
THE AZTECS Davíd Carrasco
BABYLONIA Trevor Bryce
BACTERIA Sebastian G. B. Amyes
BARTHES Jonathan Culler
THE BEATS David Sterritt
BEAUTY Roger Scruton
BESTSELLERS John Sutherland
THE BIBLE John Riches
BIBLICAL ARCHAEOLOGY Eric H. Cline
BIOGRAPHY Hermione Lee
BLACK HOLES Katherine Blundell
BLOOD Chris Cooper
THE BLUES Elijah Wald
THE BODY Chris Shilling
THE BOOK OF MORMON Terryl Givens
BORDERS Alexander C. Diener and Joshua Hagen
THE BRAIN Michael O'Shea
THE BRICS Andrew F. Cooper
THE BRITISH CONSTITUTION Martin Loughlin
THE BRITISH EMPIRE Ashley Jackson
BRITISH POLITICS Anthony Wright
BUDDHA Michael Carrithers
BUDDHISM Damien Keown
BUDDHIST ETHICS Damien Keown
BYZANTIUM Peter Sarris
CALVINISM Jon Balserak
CANCER Nicholas James
CAPITALISM James Fulcher
CATHOLICISM Gerald O'Collins
CAUSATION Stephen Mumford and Rani Lill Anjum
THE CELL Terence Allen and Graham Cowling
THE CELTS Barry Cunliffe
CHAOS Leonard Smith
CHEMISTRY Peter Atkins
CHILD PSYCHOLOGY Usha Goswami
CHILDREN'S LITERATURE Kimberley Reynolds
CHINESE LITERATURE Sabina Knight
CHOICE THEORY Michael Allingham
CHRISTIAN ART Beth Williamson
CHRISTIAN ETHICS D. Stephen Long
CHRISTIANITY Linda Woodhead
CITIZENSHIP Richard Bellamy
CIVIL ENGINEERING David Muir Wood
CLASSICAL LITERATURE William Allan
CLASSICAL MYTHOLOGY Helen Morales
CLASSICS Mary Beard and John Henderson
CLAUSEWITZ Michael Howard
CLIMATE Mark Maslin
CLIMATE CHANGE Mark Maslin
COGNITIVE NEUROSCIENCE Richard Passingham
THE COLD WAR Robert McMahon
COLONIAL AMERICA Alan Taylor
COLONIAL LATIN AMERICAN LITERATURE Rolena Adorno
COMBINATORICS Robin Wilson
COMEDY Matthew Bevis
COMMUNISM Leslie Holmes
COMPLEXITY John H. Holland
THE COMPUTER Darrel Ince
COMPUTER SCIENCE Subrata Dasgupta
CONFUCIANISM Daniel K. Gardner
THE CONQUISTADORS Matthew Restall and Felipe Fernández-Armesto
CONSCIENCE Paul Strohm
CONSCIOUSNESS Susan Blackmore
CONTEMPORARY ART Julian Stallabrass
CONTEMPORARY FICTION Robert Eaglestone
CONTINENTAL PHILOSOPHY Simon Critchley
COPERNICUS Owen Gingerich
CORAL REEFS Charles Sheppard
CORPORATE SOCIAL RESPONSIBILITY Jeremy Moon
CORRUPTION Leslie Holmes
COSMOLOGY Peter Coles
CRIME FICTION Richard Bradford
CRIMINAL JUSTICE Julian V. Roberts
CRITICAL THEORY Stephen Eric Bronner

THE CRUSADES Christopher Tyerman
CRYPTOGRAPHY Fred Piper and Sean Murphy
CRYSTALLOGRAPHY A. M. Glazer
THE CULTURAL REVOLUTION Richard Curt Kraus
DADA AND SURREALISM David Hopkins
DANTE Peter Hainsworth and David Robey
DARWIN Jonathan Howard
THE DEAD SEA SCROLLS Timothy Lim
DECOLONIZATION Dane Kennedy
DEMOCRACY Bernard Crick
DERRIDA Simon Glendinning
DESCARTES Tom Sorell
DESERTS Nick Middleton
DESIGN John Heskett
DEVELOPMENTAL BIOLOGY Lewis Wolpert
THE DEVIL Darren Oldridge
DIASPORA Kevin Kenny
DICTIONARIES Lynda Mugglestone
DINOSAURS David Norman
DIPLOMACY Joseph M. Siracusa
DOCUMENTARY FILM Patricia Aufderheide
DREAMING J. Allan Hobson
DRUGS Les Iversen
DRUIDS Barry Cunliffe
EARLY MUSIC Thomas Forrest Kelly
THE EARTH Martin Redfern
EARTH SYSTEM SCIENCE Tim Lenton
ECONOMICS Partha Dasgupta
EDUCATION Gary Thomas
EGYPTIAN MYTH Geraldine Pinch
EIGHTEENTH-CENTURY BRITAIN Paul Langford
THE ELEMENTS Philip Ball
EMOTION Dylan Evans
EMPIRE Stephen Howe
ENGELS Terrell Carver
ENGINEERING David Blockley
ENGLISH LITERATURE Jonathan Bate
THE ENLIGHTENMENT John Robertson
ENTREPRENEURSHIP Paul Westhead and Mike Wright
ENVIRONMENTAL ECONOMICS Stephen Smith
ENVIRONMENTAL POLITICS Andrew Dobson
EPICUREANISM Catherine Wilson
EPIDEMIOLOGY Rodolfo Saracci
ETHICS Simon Blackburn
ETHNOMUSICOLOGY Timothy Rice
THE ETRUSCANS Christopher Smith
EUGENICS Philippa Levine
THE EUROPEAN UNION John Pinder and Simon Usherwood
EVOLUTION Brian and Deborah Charlesworth
EXISTENTIALISM Thomas Flynn
EXPLORATION Stewart A. Weaver
THE EYE Michael Land
FAMILY LAW Jonathan Herring
FASCISM Kevin Passmore
FASHION Rebecca Arnold
FEMINISM Margaret Walters
FILM Michael Wood
FILM MUSIC Kathryn Kalinak
THE FIRST WORLD WAR Michael Howard
FOLK MUSIC Mark Slobin
FOOD John Krebs
FORENSIC PSYCHOLOGY David Canter
FORENSIC SCIENCE Jim Fraser
FORESTS Jaboury Ghazoul
FOSSILS Keith Thomson
FOUCAULT Gary Gutting
THE FOUNDING FATHERS R. B. Bernstein
FRACTALS Kenneth Falconer
FREE SPEECH Nigel Warburton
FREE WILL Thomas Pink
FRENCH LITERATURE John D. Lyons
THE FRENCH REVOLUTION William Doyle
FREUD Anthony Storr
FUNDAMENTALISM Malise Ruthven
FUNGI Nicholas P. Money
GALAXIES John Gribbin
GALILEO Stillman Drake
GAME THEORY Ken Binmore
GANDHI Bhikhu Parekh
GENES Jonathan Slack
GENIUS Andrew Robinson
GEOGRAPHY John Matthews and David Herbert

GEOPOLITICS Klaus Dodds
GERMAN LITERATURE Nicholas Boyle
GERMAN PHILOSOPHY Andrew Bowie
GLOBAL CATASTROPHES Bill McGuire
GLOBAL ECONOMIC HISTORY Robert C. Allen
GLOBALIZATION Manfred Steger
GOD John Bowker
GOETHE Ritchie Robertson
THE GOTHIC Nick Groom
GOVERNANCE Mark Bevir
THE GREAT DEPRESSION AND THE NEW DEAL Eric Rauchway
HABERMAS James Gordon Finlayson
HAPPINESS Daniel M. Haybron
THE HARLEM RENAISSANCE Cheryl A. Wall
THE HEBREW BIBLE AS LITERATURE Tod Linafelt
HEGEL Peter Singer
HEIDEGGER Michael Inwood
HERMENEUTICS Jens Zimmermann
HERODOTUS Jennifer T. Roberts
HIEROGLYPHS Penelope Wilson
HINDUISM Kim Knott
HISTORY John H. Arnold
THE HISTORY OF ASTRONOMY Michael Hoskin
THE HISTORY OF CHEMISTRY William H. Brock
THE HISTORY OF LIFE Michael Benton
THE HISTORY OF MATHEMATICS Jacqueline Stedall
THE HISTORY OF MEDICINE William Bynum
THE HISTORY OF TIME Leofranc Holford-Strevens
HIV AND AIDS Alan Whiteside
HOBBES Richard Tuck
HOLLYWOOD Peter Decherney
HOME Michael Allen Fox
HORMONES Martin Luck
HUMAN ANATOMY Leslie Klenerman
HUMAN EVOLUTION Bernard Wood
HUMAN RIGHTS Andrew Clapham
HUMANISM Stephen Law
HUME A. J. Ayer
HUMOUR Noël Carroll
THE ICE AGE Jamie Woodward
IDEOLOGY Michael Freeden
INDIAN CINEMA Ashish Rajadhyaksha
INDIAN PHILOSOPHY Sue Hamilton
INFECTIOUS DISEASE Marta L. Wayne and Benjamin M. Bolker
INFORMATION Luciano Floridi
INNOVATION Mark Dodgson and David Gann
INTELLIGENCE Ian J. Deary
INTERNATIONAL LAW Vaughan Lowe
INTERNATIONAL MIGRATION Khalid Koser
INTERNATIONAL RELATIONS Paul Wilkinson
INTERNATIONAL SECURITY Christopher S. Browning
IRAN Ali M. Ansari
ISLAM Malise Ruthven
ISLAMIC HISTORY Adam Silverstein
ISOTOPES Rob Ellam
ITALIAN LITERATURE Peter Hainsworth and David Robey
JESUS Richard Bauckham
JOURNALISM Ian Hargreaves
JUDAISM Norman Solomon
JUNG Anthony Stevens
KABBALAH Joseph Dan
KAFKA Ritchie Robertson
KANT Roger Scruton
KEYNES Robert Skidelsky
KIERKEGAARD Patrick Gardiner
KNOWLEDGE Jennifer Nagel
THE KORAN Michael Cook
LANDSCAPE ARCHITECTURE Ian H. Thompson
LANDSCAPES AND GEOMORPHOLOGY Andrew Goudie and Heather Viles
LANGUAGES Stephen R. Anderson
LATE ANTIQUITY Gillian Clark
LAW Raymond Wacks
THE LAWS OF THERMODYNAMICS Peter Atkins
LEADERSHIP Keith Grint
LEARNING Mark Haselgrove
LEIBNIZ Maria Rosa Antognazza
LIBERALISM Michael Freeden
LIGHT Ian Walmsley

LINCOLN Allen C. Guelzo
LINGUISTICS Peter Matthews
LITERARY THEORY Jonathan Culler
LOCKE John Dunn
LOGIC Graham Priest
LOVE Ronald de Sousa
MACHIAVELLI Quentin Skinner
MADNESS Andrew Scull
MAGIC Owen Davies
MAGNA CARTA Nicholas Vincent
MAGNETISM Stephen Blundell
MALTHUS Donald Winch
MANAGEMENT John Hendry
MAO Delia Davin
MARINE BIOLOGY Philip V. Mladenov
THE MARQUIS DE SADE John Phillips
MARTIN LUTHER Scott H. Hendrix
MARTYRDOM Jolyon Mitchell
MARX Peter Singer
MATERIALS Christopher Hall
MATHEMATICS Timothy Gowers
THE MEANING OF LIFE Terry Eagleton
MEASUREMENT David Hand
MEDICAL ETHICS Tony Hope
MEDICAL LAW Charles Foster
MEDIEVAL BRITAIN John Gillingham and Ralph A. Griffiths
MEDIEVAL LITERATURE Elaine Treharne
MEDIEVAL PHILOSOPHY John Marenbon
MEMORY Jonathan K. Foster
METAPHYSICS Stephen Mumford
THE MEXICAN REVOLUTION Alan Knight
MICHAEL FARADAY Frank A. J. L. James
MICROBIOLOGY Nicholas P. Money
MICROECONOMICS Avinash Dixit
MICROSCOPY Terence Allen
THE MIDDLE AGES Miri Rubin
MILITARY JUSTICE Eugene R. Fidell
MINERALS David Vaughan
MODERN ART David Cottington
MODERN CHINA Rana Mitter
MODERN DRAMA Kirsten E. Shepherd-Barr
MODERN FRANCE Vanessa R. Schwartz
MODERN IRELAND Senia Pašeta
MODERN ITALY Anna Cento Bull
MODERN JAPAN Christopher Goto-Jones
MODERN LATIN AMERICAN LITERATURE Roberto González Echevarría
MODERN WAR Richard English
MODERNISM Christopher Butler
MOLECULAR BIOLOGY Aysha Divan and Janice A. Royds
MOLECULES Philip Ball
THE MONGOLS Morris Rossabi
MOONS David A. Rothery
MORMONISM Richard Lyman Bushman
MOUNTAINS Martin F. Price
MUHAMMAD Jonathan A. C. Brown
MULTICULTURALISM Ali Rattansi
MUSIC Nicholas Cook
MYTH Robert A. Segal
THE NAPOLEONIC WARS Mike Rapport
NATIONALISM Steven Grosby
NELSON MANDELA Elleke Boehmer
NEOLIBERALISM Manfred Steger and Ravi Roy
NETWORKS Guido Caldarelli and Michele Catanzaro
THE NEW TESTAMENT Luke Timothy Johnson
THE NEW TESTAMENT AS LITERATURE Kyle Keefer
NEWTON Robert Iliffe
NIETZSCHE Michael Tanner
NINETEENTH-CENTURY BRITAIN Christopher Harvie and H. C. G. Matthew
THE NORMAN CONQUEST George Garnett
NORTH AMERICAN INDIANS Theda Perdue and Michael D. Green
NORTHERN IRELAND Marc Mulholland
NOTHING Frank Close
NUCLEAR PHYSICS Frank Close
NUCLEAR POWER Maxwell Irvine
NUCLEAR WEAPONS Joseph M. Siracusa
NUMBERS Peter M. Higgins
NUTRITION David A. Bender
OBJECTIVITY Stephen Gaukroger

THE OLD TESTAMENT
Michael D. Coogan
THE ORCHESTRA D. Kern Holoman
ORGANIZATIONS Mary Jo Hatch
PANDEMICS Christian W. McMillen
PAGANISM Owen Davies
THE PALESTINIAN-ISRAELI CONFLICT Martin Bunton
PARTICLE PHYSICS Frank Close
PAUL E. P. Sanders
PEACE Oliver P. Richmond
PENTECOSTALISM William K. Kay
THE PERIODIC TABLE Eric R. Scerri
PHILOSOPHY Edward Craig
PHILOSOPHY IN THE ISLAMIC WORLD Peter Adamson
PHILOSOPHY OF LAW
Raymond Wacks
PHILOSOPHY OF SCIENCE
Samir Okasha
PHOTOGRAPHY Steve Edwards
PHYSICAL CHEMISTRY Peter Atkins
PILGRIMAGE Ian Reader
PLAGUE Paul Slack
PLANETS David A. Rothery
PLANTS Timothy Walker
PLATE TECTONICS Peter Molnar
PLATO Julia Annas
POLITICAL PHILOSOPHY David Miller
POLITICS Kenneth Minogue
POSTCOLONIALISM Robert Young
POSTMODERNISM Christopher Butler
POSTSTRUCTURALISM
Catherine Belsey
PREHISTORY Chris Gosden
PRESOCRATIC PHILOSOPHY
Catherine Osborne
PRIVACY Raymond Wacks
PROBABILITY John Haigh
PROGRESSIVISM Walter Nugent
PROTESTANTISM Mark A. Noll
PSYCHIATRY Tom Burns
PSYCHOANALYSIS Daniel Pick
PSYCHOLOGY Gillian Butler and Freda McManus
PSYCHOTHERAPY Tom Burns and Eva Burns-Lundgren
PUBLIC ADMINISTRATION
Stella Z. Theodoulou and Ravi K. Roy
PUBLIC HEALTH Virginia Berridge
PURITANISM Francis J. Bremer
THE QUAKERS Pink Dandelion
QUANTUM THEORY
John Polkinghorne
RACISM Ali Rattansi
RADIOACTIVITY Claudio Tuniz
RASTAFARI Ennis B. Edmonds
THE REAGAN REVOLUTION Gil Troy
REALITY Jan Westerhoff
THE REFORMATION Peter Marshall
RELATIVITY Russell Stannard
RELIGION IN AMERICA Timothy Beal
THE RENAISSANCE Jerry Brotton
RENAISSANCE ART
Geraldine A. Johnson
REVOLUTIONS Jack A. Goldstone
RHETORIC Richard Toye
RISK Baruch Fischhoff and John Kadvany
RITUAL Barry Stephenson
RIVERS Nick Middleton
ROBOTICS Alan Winfield
ROMAN BRITAIN Peter Salway
THE ROMAN EMPIRE
Christopher Kelly
THE ROMAN REPUBLIC
David M. Gwynn
ROMANTICISM Michael Ferber
ROUSSEAU Robert Wokler
RUSSELL A. C. Grayling
RUSSIAN HISTORY Geoffrey Hosking
RUSSIAN LITERATURE Catriona Kelly
THE RUSSIAN REVOLUTION
S. A. Smith
SAVANNAS Peter A. Furley
SCHIZOPHRENIA Chris Frith and Eve Johnstone
SCHOPENHAUER
Christopher Janaway
SCIENCE AND RELIGION
Thomas Dixon
SCIENCE FICTION David Seed
THE SCIENTIFIC REVOLUTION
Lawrence M. Principe
SCOTLAND Rab Houston
SEXUALITY Véronique Mottier
SHAKESPEARE'S COMEDIES
Bart van Es
SIKHISM Eleanor Nesbitt
THE SILK ROAD James A. Millward
SLANG Jonathon Green
SLEEP Steven W. Lockley and Russell G. Foster

SOCIAL AND CULTURAL ANTHROPOLOGY John Monaghan and Peter Just
SOCIAL PSYCHOLOGY Richard J. Crisp
SOCIAL WORK Sally Holland and Jonathan Scourfield
SOCIALISM Michael Newman
SOCIOLINGUISTICS John Edwards
SOCIOLOGY Steve Bruce
SOCRATES C. C. W. Taylor
SOUND Mike Goldsmith
THE SOVIET UNION Stephen Lovell
THE SPANISH CIVIL WAR Helen Graham
SPANISH LITERATURE Jo Labanyi
SPINOZA Roger Scruton
SPIRITUALITY Philip Sheldrake
SPORT Mike Cronin
STARS Andrew King
STATISTICS David J. Hand
STEM CELLS Jonathan Slack
STRUCTURAL ENGINEERING David Blockley
STUART BRITAIN John Morrill
SUPERCONDUCTIVITY Stephen Blundell
SYMMETRY Ian Stewart
TAXATION Stephen Smith
TEETH Peter S. Ungar
TELESCOPES Geoff Cottrell
TERRORISM Charles Townshend
THEATRE Marvin Carlson
THEOLOGY David F. Ford
THOMAS AQUINAS Fergus Kerr
THOUGHT Tim Bayne
TIBETAN BUDDHISM Matthew T. Kapstein
TOCQUEVILLE Harvey C. Mansfield
TRAGEDY Adrian Poole
TRANSLATION Matthew Reynolds
THE TROJAN WAR Eric H. Cline
TRUST Katherine Hawley
THE TUDORS John Guy
TWENTIETH-CENTURY BRITAIN Kenneth O. Morgan
THE UNITED NATIONS Jussi M. Hanhimäki
THE U.S. CONGRESS Donald A. Ritchie
THE U.S. SUPREME COURT Linda Greenhouse
UTOPIANISM Lyman Tower Sargent
THE VIKINGS Julian Richards
VIRUSES Dorothy H. Crawford
WAR AND TECHNOLOGY Alex Roland
WATER John Finney
THE WELFARE STATE David Garland
WILLIAM SHAKESPEARE Stanley Wells
WITCHCRAFT Malcolm Gaskill
WITTGENSTEIN A. C. Grayling
WORK Stephen Fineman
WORLD MUSIC Philip Bohlman
THE WORLD TRADE ORGANIZATION Amrita Narlikar
WORLD WAR II Gerhard L. Weinberg
WRITING AND SCRIPT Andrew Robinson
ZIONISM Michael Stanislawski

Available soon:

ROCKS Jan Zalasiewicz
BANKING John Goddard and John O. S. Wilson
ASIAN AMERICAN HISTORY Madeline Y. Hsu
PANDEMICS Christian W. McMillen
ZIONISM Michael Stanislawski

For more information visit our website

www.oup.com/vsi/

Geoff Cottrell

TELESCOPES

A Very Short Introduction

Great Clarendon Street, Oxford, OX2 6DP,
United Kingdom

Oxford University Press is a department of the University of Oxford.
It furthers the University's objective of excellence in research, scholarship,
and education by publishing worldwide. Oxford is a registered trade mark of
Oxford University Press in the UK and in certain other countries

First edition published in 2016

Impression: 1

Published in the United States of America by Oxford University Press
198 Madison Avenue, New York, NY 10016, United States of America

British Library Cataloguing in Publication Data
Data available

Library of Congress Control Number: 2016946825

ISBN 978–0–19–874586–0

Printed in Great Britain by
Ashford Colour Press Ltd, Gosport, Hampshire

Contents

List of illustrations xiii

1 Introduction 1

2 Grasping light 6

3 Through the looking glass 19

4 Windows in the sky 39

5 Instruments of light 47

6 A mirror held up to nature 63

7 The radio sky 77

8 Space telescopes 96

9 The next telescopes 122

Further reading 139

Publisher's acknowledgements 141

Index 143

List of illustrations

1 Timeline of critical points in the history of the telescope **3**

2 A simple pinhole camera; the human eye **7**

3 (a) Thomas Young's double slit experiment showing bright and dark interference fringes. (b) Constructive, and (c) destructive interference of two waves **10**

4 Telescope optical elements: (a) lens, (b) mirror **14**

5 A two-lens Keplerian refracting telescope with a large objective lens (focal length *F*), and a smaller eyepiece with shorter focal length F_e. The image is inverted **15**

6 Reflecting telescopes **23**

7 Telescope mounts: (a) an altitude-azimuth mount, and (b) an equatorial mount; (c) the Japanese 8.2 m Subaru telescope, on a modern computer-controlled altitude-azimuth mount **26**
(c) © Subaru Telescope, NAOJ

8 Achromatic doublet lens **28**

9 The 100-inch Hooker Telescope on Mount Wilson, used by Edwin Hubble to measure the distances of galaxies **33**
© Science & Society Picture Library/ Getty Images

10 The electromagnetic spectrum **41**

11 The most perfect blackbody spectrum found in nature: the cosmic microwave background (CMB), measured by the *Cosmic Background Explorer* (COBE) satellite in 1989 **43**

12 A one-hour exposure of the Andromeda galaxy with its two dwarf companion galaxies, middle left and lower right, juxtaposed with a 0.01 second exposure of the Moon, to scale **48**
Photo by Geoff Cottrell

13 A transmission diffraction grating illuminated with blue, and red light **60**

14 The collecting area of optical telescopes doubles every thirty years **64**

15 Bird's eye view of the Keck II Telescope's primary mirror, showing the hexagonal segments and a man's reflection, for scale **66**
© Laurie Hatch

16 Schematic of an adaptive optics system, to correct wavefront distortions **71**

17 The radial velocity of the star 51 Pegasi b, showing the periodic trace indicating the presence of an exoplanet **75**
This plot was retrieved from the Exoplanet Orbit Database and the Exoplanet Data Explorer at exoplanets.org, maintained by Dr Jason Wright, Dr Geoff Marcy, and the California Planet Survey consortium

18 The 250-foot dish of the Lovell Radio Telescope **79**
Ian Morison, University of Manchester

19 (a) A two-antenna interferometer with baseline *D*; (b) synthesizing the aperture of a giant dish of focal length *F* **81**

20 The Karl G. Jansky Very Large Array (JVLA) radio telescope in New Mexico **86**
© Joe McNally/Getty Images

21 The very large array 6 cm map of one of the brightest radio sources: the radio galaxy Cygnus A (3C405) **87**
Image courtesy of National Radio Astronomy Observatory/AUI

22 HL Tauri: a solar system in the making, a young star and its protoplanetary system observed with the ALMA telescope, compared with the size of the Solar System **94**
ALMA (ESO/NAOJ/NRAO)

23 Crescent-shaped images of the Sun during a solar eclipse cast by gaps between tree leaves **99**
Image © Justin Soffer, flickr

24 (a) A grazing incidence (Wolter) X-ray telescope. Incoming X-rays bounce off the inclined mirror surfaces to the focus; (b) telescope mirrors on the Chandra X-ray Observatory **102**
(b) NASA/JPL-Caltech

25 The Hubble Space Telescope is the size of a bus **105**
NASA & ESA

26 A natural telescope. The gravitational lens of a massive galaxy or cluster of galaxies bends light from a distant object, and generates multiple images of the object **107**

27 The *Smiley Face*: the gravitational lensing of distant galaxies (curved arcs) by the mass of a nearer cluster of galaxies **108**
NASA/ESA/JPL-Caltech

28 The five Lagrangian points **111**

29 Three cosmic microwave background (CMB) satellites, showing the improvement in resolution of the CMB temperature fluctuations measured in the same area of sky **115**
NASA/JPL-Caltech/ESA

30 The Crab nebula at different wavelengths **120**
X-Ray: NASA/CXC/SAO; Optical: Palomar Observatory; Infrared: 2MASS/UMass/IPAC-Caltech/NASA/NSF; Radio: NRAO/AUI/NSF

31 Artist's impression of the European Extremely Large Telescope (E-ELT) **124**
ESO/L. Calçada

32 The 6.5-metre James Webb Space Telescope mirror sitting on its sunshield **130**
NASA

33 An Imaging Atmospheric Cherenkov Telescope **131**
From *Reflecting Surfaces of Novel Cherenkov Telescopes*, 19 May 2011, by Rodolfo Canestrari. SPIE Newsroom. DOI: 10.1117/2.1201105.003727

Chapter 1
Introduction

The *Oxford English Dictionary* defines a *telescope* as a 'far-seeing' optical instrument, a device used to make distant objects appear to be nearer, and larger. While the definition is true enough, it fails to capture the scope and significance of modern telescopes. The tremendous sensitivity and range of wavelengths with which telescopes now operate has transformed our knowledge of the Universe, by making the invisible visible.

This *Very Short Introduction* maps a journey from Galileo Galilei's telescope (with which he made observations, jotted down on a few pages of his notebook, that shattered a whole world view) to a telescope that will, in a few years, generate so much data that it will far exceed the current capacity of the Internet. This book is *not* a history of telescopes, but there is some history. It is about the science of telescopes: what do they do, how they work, what we have learned by using them, and where they are going next.

Telescopes exist to manipulate a substance—light. To understand the workings of even the simplest spyglass we will need to understand the rudiments of the science of light: what it is, how it propagates, when it can be described as a wave and when a particle, and how it interacts with matter. Visible light is only a small part of a much bigger spectrum of electromagnetic radiation, and while the visible portion is very important, telescopes cover a wide range of other wavelengths where some very interesting things happen. Working at these other wavelengths brings its own challenges. For example, high-energy light, like X-rays and gamma (γ) rays, simply smashes its way through matter, as is evident from medical X-rays. It follows that the familiar mirrors and lenses used to collect and focus visible light simply won't work with light of this energy, and other ways of making images have had to be found.

The Earth's atmosphere plays an important part in the story. Not only does it make stars twinkle, limiting the amount of detail that telescopes can see, but it also blocks out a great swathe of electromagnetic waves arriving from space. While we owe our existence to this naturally occurring high-factor sunscreen (high-energy light kills cells), it does mean that we have had to wait until technology enabled us to send telescopes above the atmosphere to view the blocked light. It is only with the coming of space observatories that new eyes have turned towards some of the most exotic processes taking place in the Universe.

The history of telescopes is punctuated by notable highlights, representing breakthroughs and technological changes that have influenced their development. We will dwell on some of these, what possibilities they opened up, and technical difficulties they solved (Figure 1).

The most important feature of a telescope is that it has a large aperture to collect light; it can collect more than the eye. This means that with it we can see fainter objects than with the unaided eye. The first telescopes (refractors) used glass lenses to form a magnified image. Simple glass lenses however suffer from aberrations producing image defects. Mirrors can also magnify, and are used in reflecting telescopes. Since mirrors can be made larger than lenses, they gather more light. However, early mirrors were made from bronze and were not very efficient; so refractors remained popular for many years, particularly

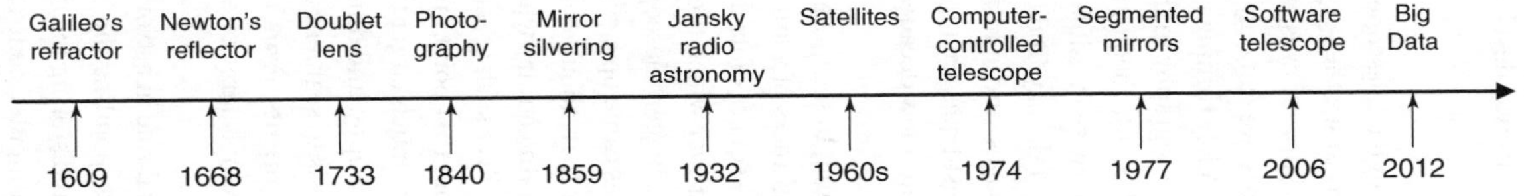

1. **Timeline of critical points in the history of the telescope.**

after the achromatic doublet lens was developed to overcome lens defects.

Gazing at a star-filled sky though a telescope is a humbling experience. But when we want to see the very faintest objects, we are limited by a feature of how our eyes work: their short 'exposure times'. In other words, we can only see objects brighter than a certain light intensity. A key turning point for telescopes came with the invention of photography in the 19th century. From then on, cameras could be attached to telescopes to make long exposures, revealing otherwise invisible objects, and to produce a permanent record of the sky. Today's electronic light detectors used for making images, and splitting the spectrum, are greatly advanced. The latest types are much more sensitive and versatile than those of only a few decades ago.

Astronomers are always pushing to see more distant, fainter objects. But the apparent brightness of a luminous object, like a star, decreases as the inverse square of its distance: the further away it is, the greater the area over which the star's light is spread. The answer is to make bigger telescopes, and capture more light. But refractor lenses came up against a hard 'end stop'; they can't be made any bigger than about a metre in diameter. A breakthrough came in the 19th century when a way was found to coat glass with a reflective silver surface, enabling large mirrors to be made for telescopes. From that point on, reflectors became the norm. Single-piece mirrors much larger than five metres in diameter are, however, heavy, floppy, and difficult to control. A way to overcome this hurdle is to segment the mirror, and use a computer to align the segments. From the 1970s on, this opened up a way to build extremely large telescopes.

Today, computers play a central role in astronomy, by actively moulding the shape of mirrors, pointing the telescope, controlling the detectors, and acquiring and storing the data. Commissioned in 1974, the first completely computer-controlled telescope was the Anglo-Australian Telescope (AAT). In radio astronomy, there is a software telescope, the Low-Frequency Array (LOFAR), which is an assembly of many cheaply produced radio aerials, spanning several countries, with no moving parts, and controlled entirely by computer. The computer defines the size of the 'beam' of the telescope, determines where it points, and produces images of the sky.

In recent years the amount of data produced by telescopes has grown exponentially, a trend that shows no signs of letting up. The next generation of telescopes, currently being built, will produce big data. And by that, we are talking about a mountain of data, anticipated to be so enormous that new data mining techniques are already being developed to sift through it to help make the scientific discoveries that are undoubtedly waiting to be found.

From Galileo's first breathtaking views of millions of hitherto invisible stars in the Milky Way, to today's powerful instruments, the first 400 years of the telescope have radically altered our view of the Universe. The latest ten-metre ground-based optical and infrared telescopes, equipped with sophisticated detectors, can image nascent galaxies, less than one billion years after the Big Bang. From space, a polychromatic view of some of the most exotic objects (including regions close to black holes and neutron stars) and energetic astrophysical processes taking place in the Universe are is being revealed over a vast range of microwave, infrared, optical, ultraviolet, X-ray, and γ-ray wavelengths. Space telescopes have been joined by ground-based radio telescopes to reveal the birth of new stars in stellar nurseries lying inside giant molecular clouds, and these instruments are now actively probing the conditions under which planetary systems form. Many telescopes are hunting for and finding extra-solar planets (exoplanets) around nearby stars. The next step along this path is the search for biomarkers, and evidence for extra-terrestrial life.

These are exciting times.

Chapter 2
Grasping light

On a clear dark night, we can see around 5,000 stars with our eyes from any point on Earth. But lifting even a simple telescope up towards our galaxy, the Milky Way, multiplies the number of stars visible a thousand times. Just what is it about telescopes that make them work like this, and so well?

Let's start with a very remarkable organ, the eye. It embodies the main features of a telescope. It has a light-collecting aperture (the iris), a lens to focus the light to an image, a detector (a two-dimensional sheet of 100 million light-sensitive photoreceptor cells on the retina), and a sophisticated information processor (the brain).

The eye evolved from light-sensitive patches on primitive forms such as the single-celled Euglena. Light sensitivity was an advantage to simple life forms, and eyes have evolved independently in different lineages in a variety of forms. In many animals, the sensitive patch evolved into a hollow, cup-shaped sheet of photoreceptor cells, capable of crudely sensing the direction of light. The eyecup became deeper and more enclosed, with a small hole at the front, projecting light on to a retina inside, operating as a pinhole camera (Figure 2(a)). In a pinhole camera, each point on the object is mapped to a unique point on the screen, forming an inverted image. The mollusc,

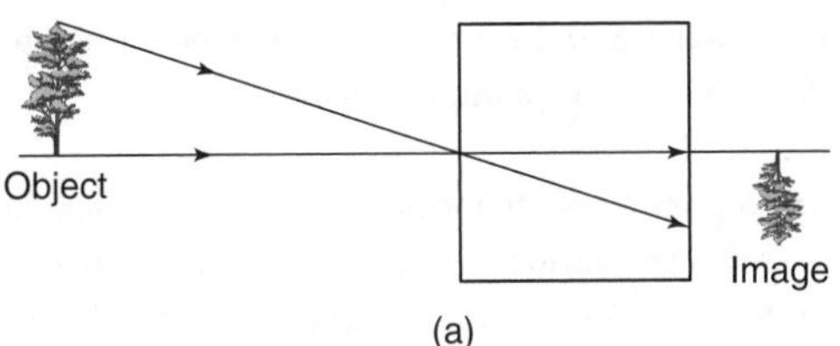

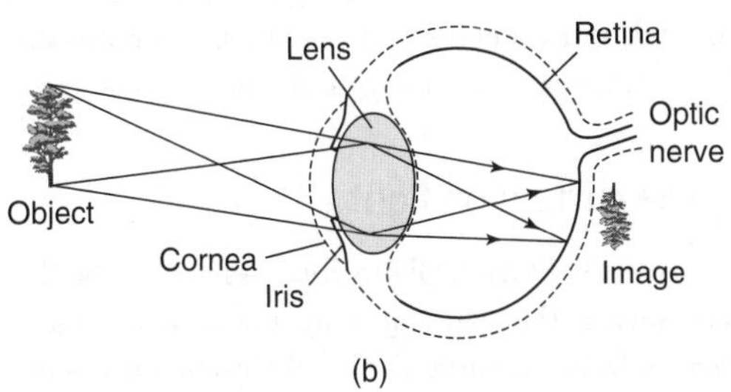

2. (a) A simple pinhole camera; (b) the human eye. The lens focuses near-parallel light to form an inverted image on the retina from where signals are sent, via the optic nerve, to the brain.

Nautilus, has eyes like this. The ability to form images was a huge step forward—but the problem was that a pinhole does not let in much light. Only rays that pass through the tiny aperture at the front are admitted, the rest are blocked. Enlarging the aperture would let in more rays, but these would come from a wider range of angles and the image would be less sharp.

The tug-of-war between sharpness and intensity was resolved when a lens evolved (perhaps a blob of clear jelly), to take the place of the pinhole. If the iris has a diameter d, the amount of light the eye gathers is proportional to the collecting area; that is, to d^2, the square of the diameter. Even with a less-than-perfect lens, the larger aperture of the eye is a big improvement. The lens bends (or refracts) the light, projecting a bright, sharp image on the retina. The human eye (Figure 2(b)) needs near-parallel rays to produce an image, and the lens muscles can be relaxed to accommodate on the near point, 25 cm away.

Telescopes also use mirrors to focus light, a pathway that has not been missed by evolution. Mirror-eyed creatures include scallops and limpets. Perhaps the most curious is the Brownsnout Spookfish. This fish has four eyes—one lensed pair facing upwards and another pair of mirror eyes facing downwards, collecting light from the murky depths of the ocean. Using both pairs of eyes, the creature can see both up and down at the same time.

Lifting the curtain on light

Our eyes collect light from all directions—from hot bodies that emit light, such as the filaments of light bulbs and stars. Light is also reflected from the surfaces of cool objects, such as planets. Light has a dual personality: sometimes it behaves like a wave, and sometimes like a particle—a photon.

Like all waves, light has a *wavelength* (usually written as λ), which is the distance between successive peaks; a *frequency*, ν (the number of vibrations a second, in units of Hertz or simply Hz); and an *amplitude*, the height of the wave. If you multiply its wavelength and frequency together, you get a speed, the *speed of light* (symbol c), which, in a vacuum (and space is the most perfect of vacuums), is 300,000 km per second, and a universal constant.

This high speed is barely noticeable in everyday life, but as a consequence of the finite speed, we see the Sun as it was eight minutes ago, and the next nearest star, Proxima Centauri, as it was four years ago. The distance light travels in one year, a light year, is almost 10 trillion or 10^{13} km (10,000,000,000,000 km), and is a measure of astronomical distance. Our nearest big galactic neighbour, the Andromeda spiral galaxy M31, is 2.5 million light years away. The light we see from its stars today started out on its journey 2.5 million years ago, long before *Homo sapiens* walked on the Earth.

In 1802, Thomas Young performed an experiment that demonstrated light's wave character (Figure 3(a)). He split a beam of light into two, using a pair of slits A and B, placing a screen behind them. The screen showed a pattern of bright and dark interference fringes, for example at C and D. A fundamental property of waves is that they can interfere with each other. If two waves are superimposed so that the peaks line up in phase, they add constructively producing a wave with twice the amplitude, as shown in Figure 3(b). If the crest of one coincides with the trough of the other, the waves interfere destructively, and they cancel (Figure 3(c)). Cancellation occurs when the waves are an exact number of *half* wavelengths out of phase. In Young's experiment the modulation of the bright and dark fringes (an interference pattern) is the key observation that revealed the wave nature of light. The distance (or path difference) from each slit to a point on the screen varies with its position on the screen. For example, the paths AC and BC differ by a whole number of wavelengths, resulting in constructive interference and a bright fringe, whereas paths AD and BD differ by a whole number of half wavelengths, producing destructive interference and a dark fringe.

To understand light's other nature, as a particle, we need to understand atoms. Atoms are made from three fundamental particles: protons (heavy, with one unit of positive charge), neutrons (heavy, but uncharged), and electrons (lightweight, with one unit of negative charge). The heavy particles are called *baryons*, and the light particles, like electrons, *leptons*. Hydrogen (H), the simplest and most abundant atom in the Universe, has a single proton in its nucleus with an electron orbiting around it. The atom is electrically neutral, the positive charge of the nucleus being balanced by the negative charge of the electron. Atoms are mostly empty space. If a hydrogen atom were enlarged to the size of London's 365 metre diameter Millennium Dome, its single

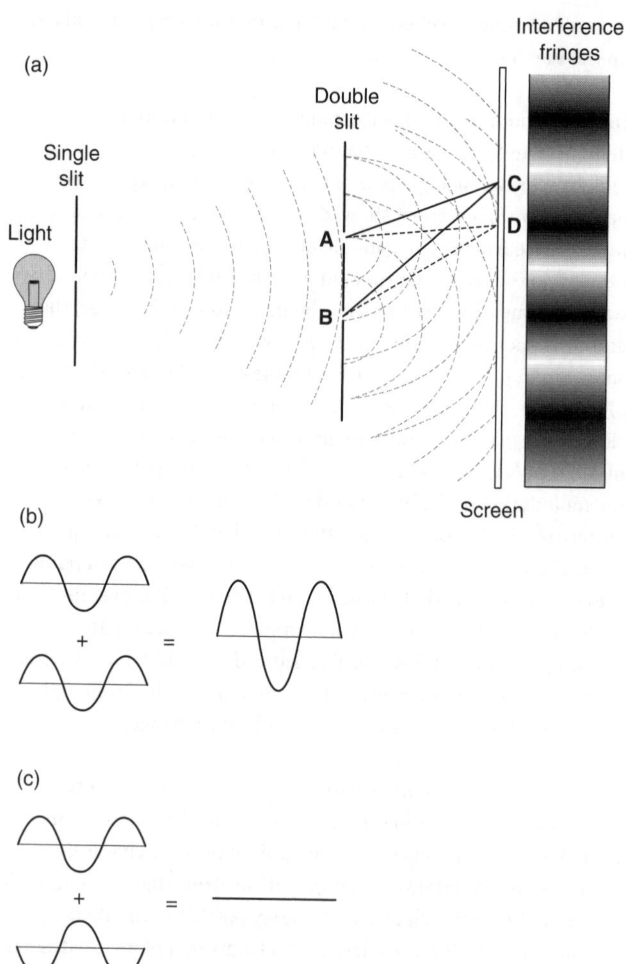

3. (a) Thomas Young's double slit experiment showing bright and dark interference fringes. (b) Constructive, and (c) destructive interference of two waves. In (a) the waves are in phase (as in AC and BC), adding to make a wave of twice the amplitude and bright fringes. In (b) the waves are exactly out of phase (as in AD and BD), and cancel making dark fringes.

proton nucleus (where most of the mass resides) would be the size of a pea.

In 1905, Albert Einstein published an explanation of a puzzling observation, the *photoelectric effect*, which had been observed some twenty years earlier by the German physicist Heinrich Hertz. When light is shone on a metal, electrons can be ejected from it. However, they are ejected only if the frequency of the light is high, even when the light source is weak. If the frequency is too low, no electrons are released, however intense the light. The wave picture of light simply could not explain this phenomenon.

Five years earlier, the German physicist Max Planck had hypothesized the quantization of radiation energy E, connecting it with its frequency with a simple expression, $E = h\nu$. The number h is known as Planck's constant. Einstein's explanation of the photoelectric effect put forward the radical idea that light is a particle, a photon, carrying a packet or quantum of energy $E = h\nu$. When a photon strikes a material, an electron absorbs the photon's energy and can be knocked out. For this to happen, the photon has to deliver a minimum amount of energy. Einstein's model explains why only higher energy photons are able to release electrons. His work led to the quantum theory of light and matter, and won him the Nobel Prize in 1921.

Photon energies are measured in electron volts (eV), a small unit that is convenient in atomic physics. An electron volt is so tiny that just by picking up a small book you expend a quintillion (a billion billion, or 10^{18}) of them. So many photons flood our eyes every second that we have the impression that light is continuous.

Each colour in the spectrum has its own wavelength and frequency; light of one colour is known as monochromatic light. For example, red light has a wavelength of 650 nanometres (one nm is 10^{-9} m), a frequency of 460 terahertz (one THz is 10^{12} Hz), and photons with energies of about two electron volts. At the blue end of the spectrum, the wavelength is 30 per cent shorter than red light, and the frequency and photon energy are correspondingly higher.

For the telescopes in this book, we need to consider three models of light. The first two assume that photon energies are much smaller than the energy sensitivity of the detectors, so that light has the appearance of being continuous. The simplest model, *geometrical optics*, is one that ignores the wave nature of light altogether and assumes that light travels between two points in a straight line. It is more accurate to say that, of the many possible paths connecting two points, light chooses the fastest (this is Pierre de Fermat's *principle of least time*). Geometrical optics applies when the wavelength of light is much smaller than the size of the equipment used.

The next model is *wave optics*, one that applies when the wavelength is comparable with the size of the equipment. If you throw a stone into a pool of smooth water, circular ripples spread out. The circles are *wavefronts*, marking the position of wave peaks. Light from a star spreads out through space in a similar way, as spherical wavefronts. The stars are so distant, that by the time their light reaches us, the wavefronts are almost flat. The Dutch scientist Christiaan Huygens proposed a principle to explain wave propagation. Every point on a wavefront is imagined to be the source of tiny spherical *wavelets* spreading out at the wave speed. At each instant, the wavelets combine to define the forward propagation of the wavefront.

The third model is *photon optics*. This comes into play when the wavelength of the light is much smaller than the size of the measuring equipment and the photon energies are much larger than the energy sensitivity of detectors. Photon optics is relevant to penetrating radiation, like X-rays and γ-rays.

Lenses, mirrors, and telescopes

The most important feature of a telescope is that it should be a good light collector. Like a bucket put out in the rain to collect raindrops, the larger the aperture of the telescope, the more photons it can gather. But a telescope also has to concentrate, or focus, the light to create an image.

When light passes through a transparent medium like glass or water, the light is slowed down by an amount defined by the medium's *refractive index*. Glass has a refractive index of 1.5, and the propagation speed of light in glass is reduced to $c/1.5 = 0.67\ c$. This slowing produces the familiar bending or refraction of light rays. A glass lens, like the one shown in Figure 4(a), works by refracting the incoming parallel light rays from a distant object, to form an image in the *focal plane*, which is at a distance from the lens known as the *focal length*. In a refracting telescope the light-collecting lens is called the objective.

How does a lens do this? Consider the plane parallel wavefronts from a distant source striking the convex lens in Figure 4(a). When the axial ray A enters the glass, its wavefronts get closer together. A second off-axis ray B strikes the surface of the lens obliquely. As soon as the leading part of B's wavefront enters the glass it is also slowed. However, the remaining part of B's wavefront still propagates at c, because it has not yet entered the glass. As the wavefront propagates through the air–glass interface, it tilts, and is refracted. On emerging from the glass, the light is again refracted and, again, continues its path through the air at speed c.

The lens converts incoming plane wavefronts to spherical ones, converging on a focus where the light interferes constructively, producing a bright spot. The lens achieves this with its time-delaying property. The central ray, A, has the shortest path.

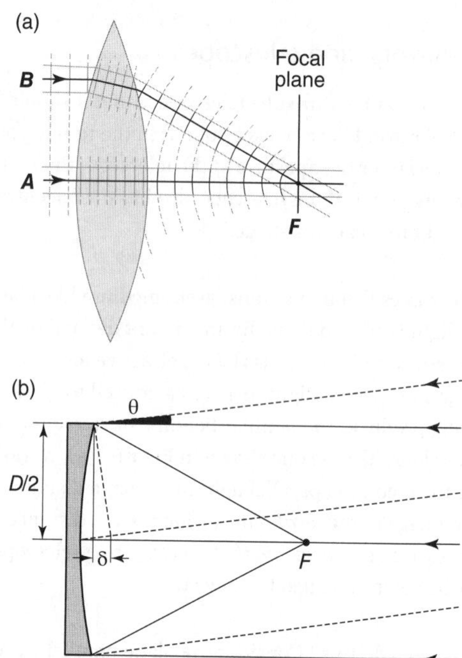

4. Telescope optical elements: (a) lens, (b) mirror. Plane parallel light is focused to the focal point *F*, in the focal plane. In (b), rays from two close-spaced stars (solid and dashed rays) are shown.

Because this ray passes through the thickest part of the lens, it is delayed more than any other ray. Ray B has a greater distance to travel to the focus and, in traversing a thinner part of the lens, is delayed less. In a properly shaped lens, each delay compensates for the exact distance a ray has to travel, ensuring that all the rays arrive in step at the focus.

The curved mirror (Figure 4(b)) can also produce an image, behaving as if it were a lens turned inside out, focussing rays by reflection, shown by the solid lines. Like a lens, a mirror converts incoming plane parallel wavefronts to spherical ones converging

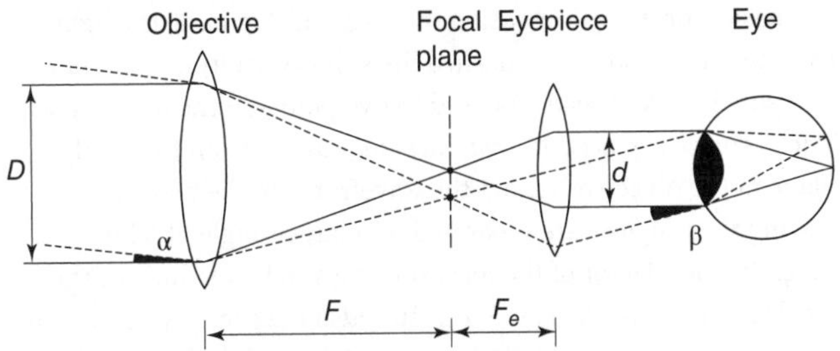

5. A two-lens Keplerian refracting telescope with a large objective lens (focal length F), and a smaller eyepiece with shorter focal length F_e. The image is inverted.

on the focus F. For this, the primary mirror must be *paraboloidal.* (A paraboloid is a 3D shape resulting from rotating a parabola around its axis of symmetry.)

It is not possible to look directly at the image produced by a lens or mirror because our eyes cannot focus on the diverging rays as they cross the image plane. To look at the image, an extra optical element, an *eyepiece*, is needed to collimate the rays, making them parallel. The simple refracting telescope of Figure 5 uses an objective (of diameter or aperture, D) and a single-lens eyepiece. Whenever a telescope is described as having a certain size D, this usually means its aperture. The light collection area is proportional to D^2, the square of the aperture. With this lens arrangement, the incoming beam is squeezed down, concentrating it into a narrow one that can pass through the pupil of the eye. The amount of light entering the eye is thereby increased by the ratio of the two areas, a factor of $(D/d)^2$. For a typical three-inch (76 mm) telescope, the light-gathering power is 600 times that of the unaided eye.

Another attribute of a telescope is *magnification.* When we use our eyes to distinguish details of distant objects (like discerning two close-spaced stars), we are trying to separate the very narrow angle between their wavefronts. The action of a telescope makes this easier for us, by increasing the angle. In Figure 5, the light from two stars (solid and dashed lines) has wavefronts separated by a small angle, α, each focused to two points in the focal plane. The eyepiece is placed its own focal length, F_e, from the focal plane of the objective so that the wavefronts of the two stars leaving the eyepiece are separated by a larger angle, β. The magnification factor of the telescope is given by the ratio of the focal lengths of the lenses, F/F_e. (It is also the same as the ratio of the objective and the pupil diameters, D/d, and the angles β/α.)

Diffraction

So far, we have assumed that light travels in straight lines, and that it is refracted as it propagates in transparent media. But its wave-like nature also makes it tend to spread out sideways, or *diffract*. The fact that lenses and mirrors cannot produce infinitely small focal points is a consequence of the finite wavelength of light. Whenever light meets an obstacle, for example an edge, or a hole in a screen, a razor-sharp shadow is not produced. Instead, the light leaks into the shadowed region. You can observe diffraction for yourself by almost touching two fingers together, making a narrow slit, and looking through it at a small bright light, like the filament of a bulb. When the slit is narrow enough, streaks of light can be seen, diffracted into lines perpendicular to the slit.

An important property of telescopes is *angular resolution*. This is defined by the angular distribution of light received from a point source at infinity. To understand how this works, let's look at what happens to monochromatic light from two close-spaced stars arriving at the focal plane of the mirror in Figure 4(b). The angle between the incoming plane wavefronts of the two stars (dashed and solid rays) is θ. When the stars are *very* close, θ is small, and diffraction blends the two images together, forming a single spot at the focus. If the angle is larger, the spot splits into two.

The question is: how small can θ be, so that two spots, and not one, can *just* be discerned?

Let's start with one star (solid rays). The star's bright central image is produced by the waves collected by the mirror, all arriving exactly in phase at F, where, as we already seen, they add constructively. The path length of each ray is the same.

When we introduce the second star, something special happens at F. This star's ray geometry is different; its rays are now slightly oblique to those of the first star. The path lengths of these rays to the focus are no longer quite equal, and this affects the way that they add up there. Angles, in astronomy, are often measured in radians, where, in the 360° of a circle, there are 2π radians. A radian is 57.3°. For small angles, the largest *path difference* (between the central and edge rays in Figure 4(b)), turns out to be:

$$\delta = \theta D / 2$$

Crucially, and this is where waves enter the picture, if the path length differences of the oblique rays are large enough, they interfere destructively at F—no light from the second star is seen there. Instead, the image of the second star has now shifted just far enough away from that of the first, so that there are images of two stars. The condition for this is when the path difference δ is a half-wavelength ($\lambda/2$) of the light, or $\delta = \theta D/2 = \lambda/2$, giving the diffraction formula:

$$\theta = \lambda / D.$$

The *smallest* angle θ (in radians) that can be resolved by a telescope is called the diffraction-limited *angular resolution*. This important formula tells us that if we want to see finer image detail with a telescope, we must use a shorter wavelength, or a larger aperture, or both.

We can immediately apply the formula to the pupil of the human eye, across which some 4,000 wavelengths of green light (wavelength 500 nm) can fit. The diffraction formula gives the angular resolution of the eye as about one arcminute (1/60th of a degree, or 1′).

Chapter 2 has introduced the main features of light, and highlighted two vitally important attributes of telescopes—the light collection area and the diffraction-limited angular resolution. In Chapter 3, these concepts will underpin the story of how telescopes have developed.

Chapter 3
Through the looking glass

> ...and so we came out [of hell] to the light of the stars.
>
> (Dante's *Inferno*)

The magnifying property of glass lenses has been known about since antiquity. Glass and spectacle-making centres have existed in Europe since medieval times, and so the essential components of a telescope have been around for centuries. However, the origin of the telescope is murky; there is apparently no record of any 'eureka' moment when it was invented.

We know that in 1608 the Dutch spectacle maker Hans Lippershey applied for a patent for a spyglass. His application was rejected on the grounds that the device was already well known and easy to copy. It seems likely that somebody had held up two lenses and discovered what the arrangement could do. The English mathematician and surveyor, Leonard Digges, is reported to have used a telescope to observe the stars as early as 1550. He wrote a popular science book describing its use which his son, Thomas, published in 1591. Unfortunately no instrument of his survives. Another English mathematician (and tutor to Sir Walter Raleigh), Thomas Harriot, observed and sketched the Moon (pre-dating Galileo Galilei's observations by four months) using a six-power telescope (his 'Dutch Truncke') in July 1609. His drawings have survived, and show a cratered lunar surface.

News of Lippershey's three-power spyglass reached Galileo in Venice, who, seeing its potential, immediately set about improving it. First he produced an eight-power instrument in 1609 (which he judiciously presented to the Venetian Senate), and then a twenty-power one. Galileo's telescope used a concave eyepiece lens, producing an upright image. Raising his telescope to the heavens, Galileo quickly made a series of revolutionary discoveries, described in his famous book: *Siderius Nuncius* (the *Starry Messenger*) of 1610.

First, his observations of the cratered and mountainous lunar surface overturned the Aristotelian notion that the Moon (like all the heavenly bodies) was a perfect sphere. Galileo discovered Jupiter's four largest (Galilean) moons, being 'towed' across the heavens by their planet. This was the first time that any astronomical body had been seen orbiting around anything other than the Earth. Galileo went on to discover that, like the Moon, the planet Venus has phases. These discoveries provided strong evidence favouring the heliocentric Copernican over the Ptolemaic model of the solar system, demoting the Earth from its supposed central position in the Universe.

Galileo's telescope had a 37 mm objective lens, eight times larger than the pupil of the eye, and so had a light-collecting power of sixty-four. This meant that he could see stars sixty-four times fainter than with unaided eyes. Since the brightness of a star decreases as the inverse square of its distance, he could therefore see stars that are on average eight times further away. When Galileo pointed his telescope at the great band of diffuse light encircling the sky, the Milky Way, the instrument revealed millions of previously unseen stars.

Imperfections

Just as the telescope was revealing the heavens to be less than mathematically perfect, the instrument itself had imperfections. The refractive index of glass varies with the wavelength of light, a dependency that gives rise to a property called *dispersion*.

Different coloured light is refracted into different angles. For example blue light is refracted more strongly than red. Dispersion is useful when you want to split light up into all the colours of the rainbow, say with a glass prism. But it can also lead to a lens defect: *chromatic aberration*. Each colour is focused at a slightly different distance from the lens. Because white light is a mixture of all colours of the spectrum, it is impossible to make sharp white-light images with a simple refractor—they are always shrouded in a blur of colour.

When making lenses or mirrors, it is much easier to figure them with spherical surfaces, rather than the ideal shapes described earlier. The ideal shapes require much careful grinding, polishing, and measurement. Spherical surfaces lead to another optical defect: *spherical aberration*. Off-axis rays have a focal length shorter than the central ones, making a fuzzy image. Even a perfectly shaped paraboloidal mirror produces yet another image defect: *coma*. This is the tendency for stellar images to become egg-shaped and grow 'tails' towards the edge of the image plane, away from the mirror axis. Coma can be corrected with additional lenses.

An important ratio in an optical system is the *focal ratio*, which is the focal length divided by the aperture (F/D), or f/number. A telescope with a high f/number enlarges the image, spreading the available light thinly over the focal plane. The terminology here comes from photography, where the focal ratio of a camera determines the *speed* of the optics. With a large focal ratio, light is more diluted and, to build up the image, a film needs a long exposure. This is a 'slow' optical system. Conversely, small f/numbers are associated with short exposures and 'fast' optics.

To minimize aberrations, 17th-century refractors were made with lenses having shallow curved surfaces, and resulted in telescopes with extremely long focal lengths. Such 'slow' telescopes are good for viewing small bright objects like stars and planets. An extreme example is Christiaan Huygens' 210-foot (64 m) long f/300 tubeless (and cumbersome) *Aerial Telescope*, which he presented to the Royal Society in 1692. The objective lens was hauled up to various heights on a tall pole, and the eyepiece (at ground level) had to be aligned with both it and the target object.

Reflectors

In 1636 a French theologian, Marin Mersenne, conceived of a telescope based on a mirror, and Isaac Newton and Laurent Cassegrain put forward designs in 1668 in 1672 respectively (Figure 6). The simplest reflecting telescope produces an image at the *prime* focus. It is generally impractical for an observer to view the image at the prime focus unless the telescope is large, like the 200-inch *Hale Telescope* (see later in this chapter), because this would block out a great deal of light.

Newton's 2-inch (25 mm) 35-power telescope is the first reflector known. In a *Newtonian* telescope, a secondary (flat) mirror is inserted into the light path, and, angled at 45° to the axis, directs the light to a side-mounted eyepiece. The telescope has a spherical solid mirror made from *speculum*, a bronze alloy (of copper and tin), polished to a shiny surface. Bronze tends to impart a yellowish hue to light, and, to whiten it, Newton added arsenic to the mix. Speculum mirrors however tarnish easily, need frequent polishing, and suffer from low reflectivity. Also the thermal expansion of this metal is larger than it is for glass, so metal mirrors are prone to distortion when observing in the cold night air. Despite these problems, speculum was the best material available for over 150 years.

In the *Cassegrain* design, the light is folded back by a secondary (convex) mirror, so that it passes through a hole in the primary. There are several variations of this basic layout. One of these, the *Gregorian*, has a concave secondary mirror, placed beyond the

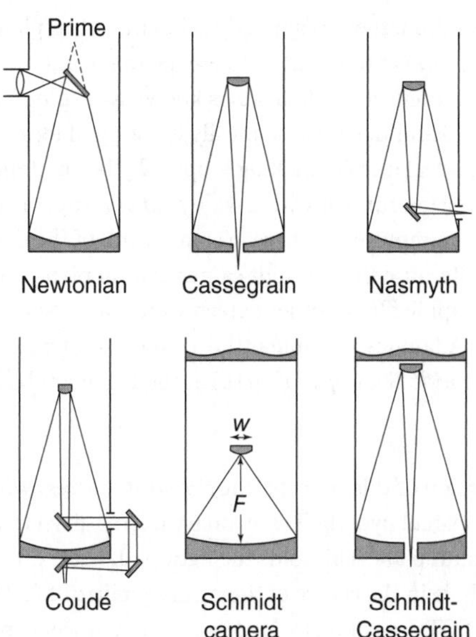

6. Reflecting telescopes. Secondary and other mirrors are used to create alternative foci. In the Schmidt telescopes, the mirror is spherical and an aspherical correcting lens is located at the front.

focus of the primary mirror, resulting in an upright image, which is viewed with a rear-mounted eyepiece. The Scottish engineer James Hall Nasmyth produced another variation, deploying a flat tertiary mirror to reflect the light sideways to a *Nasmyth* focus. Nasmyth's ingenious 20-inch (51 cm) reflector of 1849 has the eyepiece attached to one of the hollowed out shafts on which the telescope pivots. His telescope can be directed anywhere in the sky without the observer having to move. Yet another arrangement is the *Coudé* system, using flat mirrors, as shown in Figure 6, to produce alternative foci.

In 1930, Bernard Schmidt invented a different type of reflector—the *Schmidt camera*. The primary mirror is spherical, and, to correct for the aberration, there is a weakly refracting glass plate (a *Schmidt plate*) at the front. A telescope that combines refracting and reflecting elements is known as a *catadioptric* telescope. A Schmidt camera typically has a small focal ratio (a small F and a large D), making it optically fast and therefore a good survey instrument with a wide *field of view*. The field of view angle is simply w/F, where w is the width of the detector (Figure 6). Because the Schmidt camera's focal plane is curved, the photographic film or other detector must also be made that shape. A famous example is the 48-inch (1.23 m, f/2.5) *Oschin–Schmidt Telescope* (of 1948) at the Palomar observatory in California.

In an extension of this concept, the *Schmidt–Cassegrain Telescope* (SCT) has a small hyperbolic secondary mirror, fixed to the back of the Schmidt plate. This folds the light path back, directing it through a hole in the centre of the primary mirror to a Cassegrain focus. The SCT is a compact, large-aperture telescope, popular with amateur astronomers.

William Herschel designed and built several reflectors, including the then largest in the world, his *40-foot telescope* of 1789, with a 47-inch (1.2 m) mirror. Herschel eliminated the low-reflectivity flat mirror of the Newtonian, by tilting the primary mirror slightly in the tube, so that images could be viewed directly through an eyepiece placed off to one side. In 1781, while observing double stars from the garden of his house in Bath, England with his 6.2-inch (16 cm) reflector ('one with a most capital speculum'), he noticed an object shifting position from one night to the next against the background of the fixed stars. At first he thought it might be a new comet, but it was disc-like and had no tail. It turned out to be a new planet—Uranus, the first planetary discovery since ancient times. Its discovery catapulted Herschel to fame.

In the 18th century, hunting comets was an important astronomical activity. The French astronomer and avid comet hunter Charles Messier was constantly frustrated by finding fuzzy objects, which weren't comets. So, in 1781, he created a catalogue of 103 of the tiresome objects that astronomers should avoid, and on which they should not waste their time. The *Messier Catalogue* in fact contains some most interesting 'fuzzy' objects, including many galaxies such as the Andromeda galaxy (Messier's 31st object, or M31), and star-forming clouds (such as the Orion nebula, M42) in the Milky Way.

With his 20-foot telescope (47 cm mirror), Herschel, working with his sister Caroline, made extensive sky surveys, observing and documenting 2,500 new fuzzy objects that were dubbed *nebulae* (clouds). These were published in the *Catalogue of Nebulae and Clusters of Stars* of 1802, which later formed the basis of the *New General Catalogue* of 1888.

In the 1840s, William Parsons, the third Earl of Rosse, built the world's largest telescope at Birr Castle near the Bog of Allen in Ireland. It was a huge Newtonian with a 6-foot (1.8 m) mirror, known as the *Leviathan of Parsontown*. With this instrument, Rosse discovered faint nebulae with *spiral* patterns, notably the famous Whirlpool galaxy (M51). Rosse also observed and named the Crab nebula (M1).

Tracking the stars

To track the stars, a telescope has to compensate for the Earth's rotation, and follow their apparent movement on the sky. This motion is based on *sidereal* time, a system referenced to the stars, and not normal solar time. A sidereal day is four minutes shorter than a solar day. The difference arises because, in one solar day, and seen from the distant stars, the Earth has to rotate slightly *more* than one revolution owing to its orbital motion around the Sun.

Objects on the celestial sphere, an imaginary sphere projected on the sky, are located by pairs of coordinates, similar to latitude and

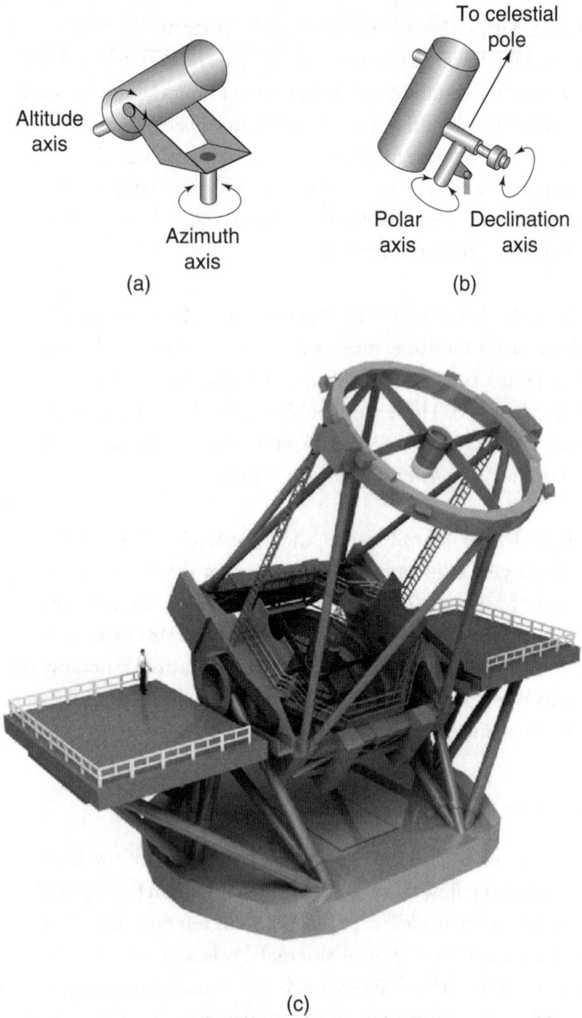

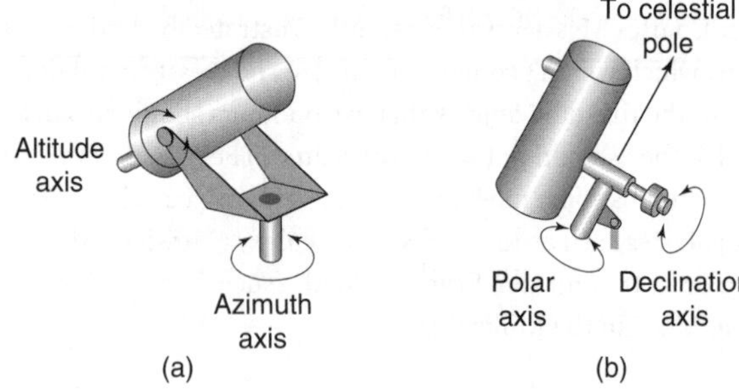

(c)

7. Telescope mounts: (a) an altitude-azimuth mount, and (b) an equatorial mount; (c) shows the Japanese 8.2 m Subaru telescope, on a modern computer-controlled altitude-azimuth mount. The telescope's prime focus is at the top, and the two large Nasmyth platforms are positioned at the IR and optical foci.

longitude on the Earth's surface, with celestial north and south poles. The north celestial pole is close to the North Star, Polaris. One coordinate (*right ascension*) is measured in hours, minutes, and seconds, and is equivalent to longitude. The other, in degrees (*declination*), is equivalent to latitude, with the celestial equator at zero declination. The great circle passing through the poles and the zenith point (directly overhead) and the nadir point (directly below) is called the meridian.

To be of any use, a telescope needs both a solid mechanical mount, and the ability to be pointed to any part of the sky with ease. In 1722, the English astronomer John Hadley developed a practical *altitude-azimuth* (altazimuth) mount (shown schematically in Figure 7(a)). An altazimuth mount can be rotated about a vertical axis (the azimuth or compass bearing), and a horizontal axis, elevated above the horizon. There are numerous examples of altazimuth mounts, one being a gun turret on a battleship. The mount is compact and can support heavy loads. In the 1960s the American John Dobson mounted a Newtonian on an altazimuth mount, producing the *Dobsonian* telescope, an arrangement popular with amateur astronomers. This portable setup enables quite large reflectors (up to about 0.5 m) to be taken to dark-sky observing sites.

However, when it comes to manually tracking stars with an altazimuth mount, things can be complicated. The telescope has to be rotated about both axes, and at different rates. This is something that can be performed easily under computer control, and most large modern telescopes, for example the *Subaru* telescope (Figure 7(c)), employ this method.

Star tracking could be greatly simplified if the azimuthal plane was tilted upwards, so that it is perpendicular to the Earth's rotational, or polar axis. This is the principle of the *equatorial* mount (Figure 7(b)). Stars can be tracked easily by rotating the telescope at a constant speed around just the polar axis. In practice, the declination axis is first set (and locked) to the declination of the target object.

One of the first equatorial mounts was Christophe Scheiner's *helioscope*, of 1638, a telescope used to project an image of the Sun on to a screen. (This reminds us that the Sun must never be looked at directly through a telescope.) In the 1820s the German optician Joseph von Fraunhofer developed the practical and popular German Equatorial Mount (GEM). He used one for his *Dorpat* (24 cm) refractor (in Tartu Observatory, Estonia), in which the polar axis was driven at a constant rate by a falling weight and clockwork mechanism.

The great refractors

In 1733, a London barrister and optical hobbyist Chester Moore Hall solved the problem of chromatic aberration. In a convex lens, red light is refracted less than blue, resulting in red rays having a longer focal length. But in a concave lens the opposite happens. Chester Moore Hall placed two lenses together, one of crown glass, and the other of flint glass (Figure 8), and found a combination that cancels out the aberration. This type of lens is known as an *achromatic doublet*.

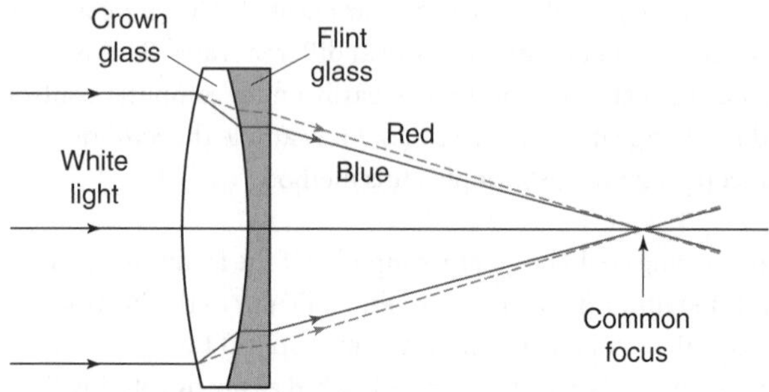

8. Achromatic doublet lens.

The English optician John Dollond further developed the doublet, patenting the concept in 1758, and making 12 cm objective glasses. By the 1820s, Fraunhofer had perfected a way of making even larger doublets, up to 24 cm in diameter. Between 1820 and 1900 many successful short-tube refractors were built using these lenses. By the end of the 19th century, refractor apertures had quadrupled in size, culminating in the largest ever to be made, the 40-inch (1 m) instrument at the Yerkes Observatory in Chicago, built by the American astronomer and entrepreneur George Ellery Hale in 1895. This instrument was to mark the high point of the refractor; a glass objective cannot be made any larger. For one thing the glass has to be free of imperfections, such as bubbles and variations of refractive index (striae) throughout its *volume*. Being edge-supported, a heavy glass lens sags unacceptably. A mirror on the other hand has only to have a near-perfect *surface*, a much less stringent requirement. But, by then, a key development for making large mirrors had taken place, and the days of the great refractor were numbered.

The breakthrough came in 1859 when Karl August von Steinhal and Léon Foucault found a way of making large mirrors by chemically depositing silver on glass. This was a key turning point; for the first time, glass could be used as a mirror substrate material. The low thermal expansion of glass meant that, from then on, large, stable, and highly reflective mirrors could be made.

The age of reflectors

Our desire to peer ever more deeply into the cosmos never diminishes. But the further away one tries to look, the harder it gets, and bigger telescopes are needed. Big telescopes capture more photons. And bigger telescopes have mirrors.

Ever since the time of Galileo (whose telescope resolved the smooth Milky Way into millions of stars), of Herschel (who catalogued thousands of nebulae), and of Parsons (who discovered spiral patterns in nebulae), astronomers have wanted to measure the size of the Universe. But measuring astronomical distance is not an easy thing to do.

At the start of the 20th century, the only distance-measuring method in astronomy was *parallax*. Hold a finger up at arm's length and close an eye. Then alternate eyes. The finger shifts position relative to the background. This is parallax. The same angular shift (although smaller) happens to nearby stars. Every six months, the Earth is on opposite sides of its orbit, and the stars that happen to be fairly close to us appear to shift, relative to the background of the much more distant 'fixed' stars. The distances of the nearer stars can be found by triangulation, with the baseline of the triangle being the diameter of the Earth's orbit (two *astronomical units* (AU), about 300 million km). In 1900, parallax could be applied to about 100 stars, all within 100 light years. The shifts of more distant ones were too small to measure.

The answer to the question of the size of the Universe was to find out the distances to the nebulae; but these were clearly more than 100 light years away. Crudely speaking, these fuzzy objects came in two varieties. Some nebulae, for example that in Orion (M42), in the Milky Way, have the appearance of a smooth luminous cloud, peppered with young and actively forming stars. Others, like the great spiral in Andromeda (M31), also present a smooth milky appearance, but with no discernible stars. Astronomers could not explain the latter group. The question then was: are the spirals simply regions of the Milky Way, or are they separate star systems lying outside it? If the latter were true, the spiral nebulae would have to be so far away that their constituent stars could not be resolved.

To measure larger distances, astronomers needed a *standard candle*. A standard candle is some type of common brightly emitting body, which has a standard intrinsic luminosity, or light output. If its *apparent* brightness can be measured, and if we know its intrinsic luminosity, then its distance can be found by applying the inverse square law.

In 1912, Henrietta Swan Leavitt was working in Harvard College Observatory measuring variable stars in the Small Magellanic Cloud, one of the Milky Way's two small, irregularly shaped neighbouring galaxies. Leavitt measured thousands of variable stars, which she knew must all be at a similar distance. Most of the stars she studied were a type known as *Cepheid variables*, which pulsate like beacons with periods lasting anything from days to months. She discovered that the more luminous ones have systematically longer periods—she had discovered a *period-luminosity relationship*. By the following year, Ejnar Hertzprung had measured geometrical distances of thirteen nearby Cepheids, enabling him to calibrate Leavitt's data in terms of distance. The upshot of all this is that if you measure the brightness *and* period of a Cepheid, you can work out its distance. The first standard candle, the Cepheid variable, was born.

In 1908, George Hale commissioned the first of three great reflectors. This was a 60-inch (1.5 m) Cassegrain telescope, located at Mount Wilson (altitude 1,740 m) near Los Angeles, California, a site chosen for its clear skies. The glass mirror blank was a gift from his father, and was ground to a paraboloid with carborundum, polished, and silver-coated. Hale's 60-inch telescope was the largest in the world, and in 1916, Harlow Shapley used it to study globular clusters. These are densely populated spherical star systems containing up to a million stars. In the Milky Way there are over one hundred clusters, spread out symmetrically; but from Earth, they are seen mostly in one-half of the sky. Shapley discovered Cepheid variables in the Milky Way's globular clusters, allowing him to estimate their distances and make a 3D map of them. Seeing that their distribution was skewed (with more towards the centre of the galaxy), he deduced that the Sun lay in the outskirts of the galactic disc.

In 1920 there was still no agreement amongst astronomers about the nagging question of the spiral nebulae. Were they made of stars or gas? Were they inside or outside the Milky Way? Observing stars in them would resolve the question, but even the 60-inch telescope could not see any. That year, the *great debate* took place, between astronomers Shapley (who argued that the spirals were local), and Heber Curtis, who claimed that they were *island universes* (or extragalactic star systems).

The 100-inch Hooker Telescope

Hale attracted funds from John Hooker and Andrew Carnegie to start building an even bigger telescope, the 100-inch (2.5 m) *Hooker Telescope* (Figure 9) on Mount Wilson. When this was completed in 1917, it would be the largest telescope in the world until 1948. It would also turn out to be the most important since Galileo's telescope.

The first job for the world's most powerful telescope was to look at the spiral nebula in Andromeda (M31), and see if any stars or star clusters could be discerned within its foggy light. The task fell to an American astronomer, Edwin Hubble, who, in 1919, painstakingly set about taking photographs. Sure enough, the big instrument revealed a powdering of extremely faint stars. At once the great debate was resolved. The Andromeda nebula was henceforth called the Andromeda galaxy. But how far away was it? In 1923, Hubble found the first of many Cepheid variables in the galaxy. By comparing them with the ones in the Magellanic Clouds, he deduced that M31 must be nearly a million light years away, placing it well outside the Milky Way. These were very important discoveries; they fundamentally changed our knowledge about the Universe, showing that the spiral nebulae are distinct external galaxies, each comparable with the Milky

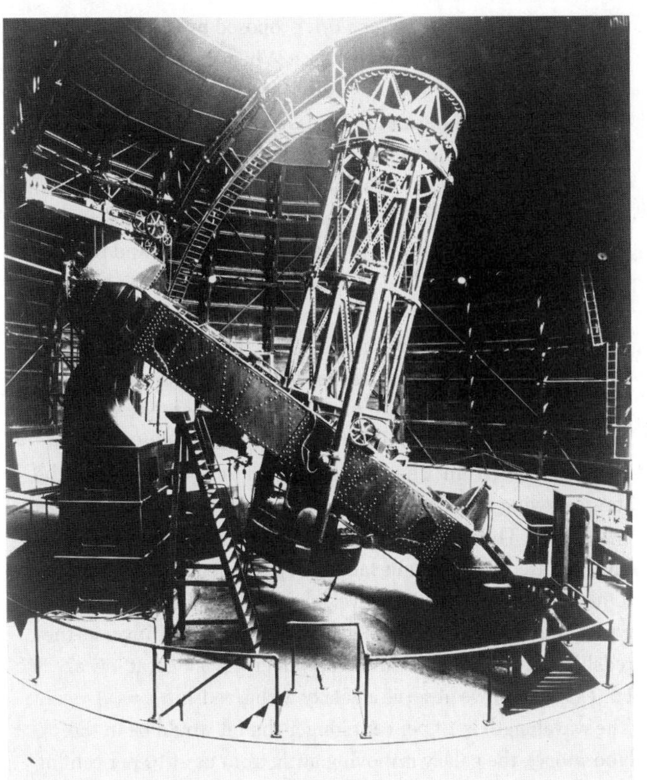

9. The 100-inch Hooker Telescope on Mount Wilson, used by Edwin Hubble to measure the distances of galaxies.

Way. Our galaxy was thenceforth relegated to just one of very many others, scattered throughout space.

Redshifts and the Hubble expansion

If you stand next to a 'Formula One' racetrack, as a racing car zooms by, the sound of its approaching engine changes from a high to a low pitch as it approaches and then recedes into the distance. This is the *Doppler shift*, proposed by the Austrian physicist Christian Doppler in 1842. When the car approaches, each successive wave peak is emitted from a position closer to you than the previous one, and so takes less time to reach you, causing the waves to bunch together. The time between the wave peaks (which is sensed as the pitch of the engine) is therefore reduced, and the pitch is raised. Conversely, when the car speeds away, the time between successive wave peaks is lengthened, and the pitch is lowered.

Light can also be Doppler shifted. When a star moves away from us, in the time of one light oscillation, the star has moved a little further away and the wavelengths of its atomic spectral lines are stretched out and shifted to the red, long-wavelength end of the spectrum. The starlight is said to be *redshifted*. If the star is moving towards us, the opposite happens and the light is *blueshifted*. The amount by which the wavelength of a spectral line is shifted ($\Delta\lambda$) from its laboratory wavelength, λ, enables us to measure an object's radial velocity, v, in the line of sight. For typical velocities of stars and galaxies in the local Universe, the redshift is $z = \Delta\lambda/\lambda$, and the radial velocity is $v = cz$ (at low z). For example, if we observe a galaxy with a redshift $z = 0.1$ (the wavelength is 10 per cent longer than it would be in the laboratory), the galaxy is moving away from us at 10 per cent of the speed of light, equivalent to a recession velocity of 30,000 km per second.

In 1912, the American astronomer Vesto Slipher measured the redshifts of galaxies. Much of the light from a galaxy comes from its hundreds of billions of stars, each with different masses, ages, chemical compositions, and radial velocities. If the galaxy is moving away from or towards us, the galaxy's spectrum shows a net red- or blueshift. Slipher found that the galaxies were moving at high radial velocities. By 1917, he had measured seventeen galaxies, and by 1925 over forty, mostly receding from us.

In some of these galaxies, Hubble and Milton Humason discovered Cepheid variable stars and so were able to measure their distances. Hubble made a plot of a galaxy's distance and its radial velocity showing that they were related: the more distant galaxies were moving away from us with higher velocities than the nearer ones. This far-reaching result is embodied in the *Hubble law*: $v = H_0R$, where v is the recession velocity of a galaxy and R its distance. The constant of proportionality, H_0, is called the Hubble constant and is a measure of the present expansion rate of the Universe.

Hubble's discovery implied that we live in a dynamically evolving, non-static Universe, a revelation that came only seventy years after the publication of Charles Darwin's revolutionary book *On the Origin of Species*. By running the cosmic clock backwards, Hubble's law predicts there must have been a time in the past (the Hubble time, $1/H_0$) at which all the galaxies coincided. This was a point of infinite density (a singularity) marking the beginning of the Universe—the *Big Bang*.

The fact that the galaxies are receding from us, in whichever direction we look, might suggest that we have a privileged position in the Universe. But we do not. The Hubble law expresses the notion that space itself is expanding, with all the galaxies embedded in it. Any observer living on any other galaxy would see exactly the same thing.

To clarify this, imagine baking a large blueberry muffin, with blueberries mixed in with the batter. At the start of the bake, the blueberries are close to each other. As cooking proceeds, the baking powder releases carbon dioxide bubbles, pushing everything apart, making the muffin expand and rise. The embedded blueberries are carried along, pushed apart by the expanding batter. Deep inside the muffin, there is no privileged blueberry; whichever one you focus on, all the other blueberries are moving away from it, in all directions. They follow a kind of 'muffin Hubble law'. The blueberries represent galaxies embedded in an expanding space (the batter). (This analogy is of course flawed because, unlike a muffin, we know of no 'edge' to the Universe.)

In 1933, the Swiss astronomer, Fritz Zwicky, used the Hooker Telescope to observe the *Coma cluster*, an aggregation of over a thousand galaxies, bound together in a clump, by gravity. He used a spectrograph to measure the radial velocities of individual galaxies. From these data, Zwicky calculated the cluster's *virial* mass (this is the gravitational mass which is needed to bind the cluster together and stop the members escaping). He compared the virial mass with mass expected from the luminosity of the member galaxies. It turned out to be many times larger, and from this he inferred that most of the matter in the Coma cluster must be dark. It was Zwicky who coined the phrase *dark matter*.

Throughout the 1930s an increasingly large question mark grew over Hubble's estimate of the distance to the galaxies. At face value, his measurements suggested that the Milky Way was unusually large, compared with neighbouring (and ostensibly similar) spiral galaxies. When Hale's telescopes were built on Mount Wilson, Los Angeles was much smaller. The city spread quickly in the 1920s, as did the light pollution, making the telescopes increasingly unusable for the faintest objects. But during World War II, the conurbation was blacked out, for fear of air raids. That meant that the Hooker Telescope could again be used, albeit briefly. In 1942, Walter Baade seized the opportunity to revisit the Andromeda galaxy, and see if he could see stars in its nucleus. He found them. He discovered two distinct stellar populations, and two types of Cepheids. The ones that Hubble had observed in M31 behaved differently from those in the Magellanic Clouds. The Hubble law was revised (with a new value for H_0); this meant that the galaxies should be roughly twice the distance previously determined.

The revision of the Hubble constant also resolved a paradox. The radioactive dating of rocks indicated that the Earth was at least 3.6 billion years old, but this did not tally with Hubble's initial value for the Hubble time which was $1/H_0 = 2$ billion years. How could the Earth be formed before the Universe came into existence? With Baade's revised distance scale, the paradox disappeared.

For Hale's last and most ambitious telescope, the 200-inch (5.1 m) *Hale Telescope*, a new dark site was found, south of Los Angeles, on mount Palomar (altitude 1712 m). The heart of every big telescope is its mirror, and this one was special. The 200-inch mirror was cast as a disc of low-expansion Pyrex glass, weighing over twenty tonnes, even with the back of the mirror ribbed, like a waffle, to reduce both its weight and the time it takes to reach ambient temperatures. Over eleven years of grinding, polishing, and testing was needed to produce a near-perfect paraboloidal surface, accurate to within two millionths of an inch (50 nm). By the time it was finished, over five tonnes of glass had been ground off the disc, and thirty tonnes of grinding compounds used up. The surface was polished and coated with a thin, vacuum-deposited film of aluminium. Aluminium is superior to silver; it doesn't tarnish and remains highly reflective at near-ultraviolet wavelengths.

A big telescope mirror must keep its shape, under the stresses and strains of gravity, as it is tilted from horizontal to near-vertical attitudes. An inadequately supported 13-tonne, 5 m diameter glass mirror will sag, ruining performance. To oppose any flexing, the Hale mirror is cradled in a massive support cell, with mechanical supports applying correcting forces. All of this complexity is needed to keep a few grams of aluminium stuck to the surface, in the right place. The mirror cell is borne on a massive yoke-shaped equatorial mount. The mirror has a focal length of 16.7 m (f/3.3). At the prime focus, there is a cage, large enough to carry an observer, moving with the telescope. There are also f/16 Cassegrain, and Coudé foci (Figure 6).

The '200-inch' played a leading role in astronomy and cosmology for at least three decades, and pioneered observations of the cosmic expansion and galaxy stellar populations—work that has greatly improved our understanding of galaxy evolution. It was also used to identify the first quasar in 1963, as well as many galaxies hosting radio sources.

We have now followed the story of optical telescopes from their beginnings, up to the threshold of the modern age of telescopes. We have seen the tug-of-war between refractors and reflectors, and how the reflector won the day. However, ground-based telescopes do not operate in isolation; the light they collect is affected by its passage though the Earth's atmosphere. In fact, our atmosphere affects much more than visible light, as we will see in Chapter 4.

Chapter 4
Windows in the sky

About half the Earth's surface is obscured by cloud, and the other half is clear. If our ancestors had evolved on a completely cloud-bound planet, they would have found it extremely difficult to grasp the concepts of space and time. Stargazers have looked out and observed the celestial motions and rhythms of the Sun, Moon, planets, and stars. This is how we acquired a sense of time and knowledge about the rhythms of the year, the seasons, and the day. It has also given us our sense of direction and the points of the compass. We taught ourselves to navigate around the globe 'by the stars', which we see at night through the thin blue line of the atmosphere, stuck to the Earth's surface by gravity. Since the atmosphere influences much of what can be seen through a telescope, we need to spend some time understanding it.

Most of the atmosphere lies within 16 km from the surface, and is composed mainly of nitrogen and oxygen. There is also water vapour, carbon dioxide, and there are traces of other gases. Further out, the air becomes thinner until it merges with outer space. In a layer 75–1,000 km high, called the *ionosphere*, neutral atoms are ionized by solar radiation and high-energy cosmic ray particles arriving from distant parts of the Universe. When an atom is ionized, an electron is stripped off, leaving behind a positively charged particle, or ion. In the process of *photoelectric absorption*, an atom absorbs a high-energy photon and ejects an electron, resulting in an ion–electron pair. When large numbers of positive ions and electrons clump together, they form a special high-temperature and electrically conducting gas, a *plasma*. A plasma that is familiar is the one inside a fluorescent lamp. In plasmas the proportion of ions and electrons is balanced so that there is virtually no net electrical charge.

Energy is needed to strip an electron off an atom, and, when the two oppositely charged particles are pulled apart, the *electric force* between them tries to pull them back. When we comb our hair or pull off a woollen sweater we experience static electric forces directly. The act of rubbing materials together separates charge: sparks fly, hair stands up, and clothes stick together. The electric force originates in an invisible *electric field* surrounding a charge, an influence that reaches out through empty space. If there is another charge nearby, there is a force between them. We are also familiar with the forces between magnets, associated with the *magnetic field*, acting through space.

Electromagnetic radiation

The electric and magnetic fields, as described earlier, are essentially static. But they don't have to be. When they vary in time, an amazing connection between them emerges. Building on the pioneering electrical experiments of Michael Faraday and others, James Clerk Maxwell published his four *Maxwell's Equations* in 1862. The equations constitute one of the most elegant and beautiful edifices in the whole of physics. They describe the *electromagnetic (EM) field*, which supports EM waves transporting energy through space at the speed of light. This is the enormous triumph of Maxwell's theory: the conclusion that light is an EM wave. An EM wave is a *transverse* wave, one where the oscillating electric and magnetic wave fields combine and march forward in lockstep, aligned perpendicularly to each other and to the direction of propagation. A key property of a transverse wave is that it can be *polarized*, something that happens when the oscillations all line up together in the same direction. Maxwell's theory predicts that EM waves are generated when charges are accelerated. This prediction was confirmed by Hertz's 1888 experiments in which electrons, oscillated in a spark transmitter, produced EM radiation from an antenna.

Beyond visible light, at both shorter and longer wavelengths, there is a vast swathe of invisible electromagnetic radiation (Figure 10). Starting from the shortest wavelengths, the wavebands are: γ-rays, X-rays, ultraviolet (UV), visible light, infrared (IR), millimetre (mm), and finally radio waves. The wavelength range is enormous, from γ-rays ($\lambda = 10^{-15}$ m, the diameter of a proton) to radio waves ($\lambda = 100$ m, the height of the Statue of Liberty, and beyond).

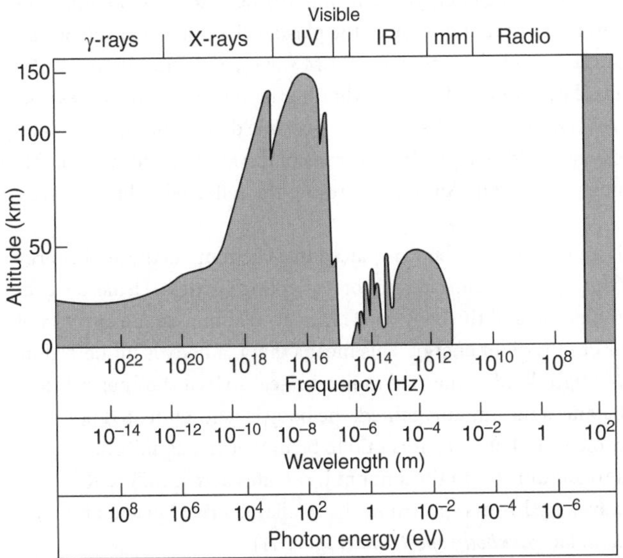

10. The electromagnetic spectrum. The shaded region is the altitude at which the atmosphere is transparent to cosmic radiation of different wavelengths.

The different wavebands merge smoothly into each other, and there are no sharp boundaries.

Heat and radiation

Heat is the random jiggling of the particles of matter; the hotter the body, the greater the jiggling. As a result of this whirling atomic dance, all hot bodies, with temperatures above absolute zero, radiate EM waves as *thermal radiation*. The temperature scale used is the *absolute temperature* scale, with absolute zero (zero degrees Kelvin (K), or –273°C) being the lowest temperature possible, and the point at which atomic motion ceases.

The light and radiant heat we see and feel from the Sun arises because the Sun is hot, powered by fusion reactions deep in its core, releasing energy by burning hydrogen to make helium. The random agitated motion of the plasma electrons near the Sun's surface produces unpolarized EM waves, with vibrations pointing in all directions. However, when sunlight or starlight shines on, say, the surface of the ocean or a cosmic dust cloud, the oscillating electric fields wiggle the electrons in the matter, scattering light in one preferred direction. As a result, the reflected light is polarized.

If an isolated, hot body is placed in a vacuum, its thermal energy (the sum of all the energies of its jiggling particles) leaks away by radiation, and the body cools. However, atoms can absorb as well as emit radiant energy. This means that if a body is immersed in the right kind of thermal radiation 'sea', so that the energy lost balances the amount gained, the body is neither heated nor cooled. In that state, it is said to be in thermal equilibrium with its surroundings, and the amount it radiates across different wavelengths—its spectrum—has a characteristic peaked form called a *blackbody spectrum* (Figure 11).

A blackbody spectrum has important properties. In 1893 the German physicist Wilhelm Wien discovered a 'displacement' law;

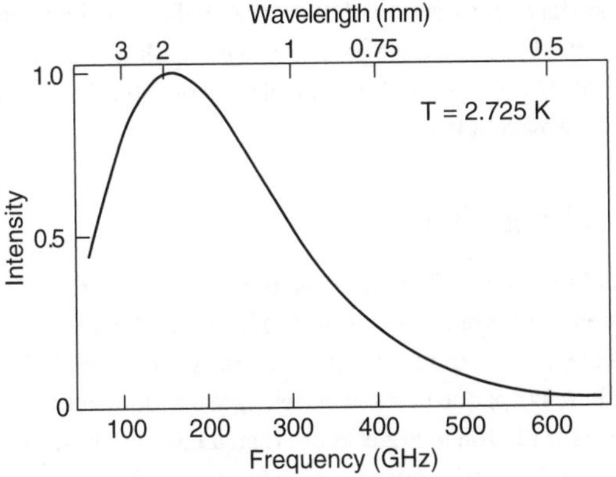

11. The most perfect blackbody spectrum found in nature: the cosmic microwave background (CMB), measured by the *Cosmic Background Explorer* (COBE) satellite in 1989. It has a temperature of 2.725 K. The measurement error bars are too small to be shown.

namely, that the wavelength of the peak of a blackbody spectrum is inversely proportional to its temperature. For example, iron at room temperature (300 K) radiates because it is hot, but we can't see its radiation because the peak of its spectrum is in the infrared (IR), wavelengths to which our eyes are not sensitive. If the metal is heated to 1,500 K, the peak of its spectrum shifts to shorter wavelengths, which becomes visible as an orange-red glow. If the wavelength of the peak of a blackbody spectrum can be measured, Wien's law can be used to estimate the temperature.

The Sun, like all stars, is a hot plasma ball of glowing gas. The surface temperature of our star is 5,800 K, and its blackbody spectrum peaks in the middle of our visual range. It is no accident that our eyes have evolved to be sensitive to this wavelength. Stars have different temperatures and so their spectra peak at different wavelengths. For example, in the constellation of Orion, Betelgeuse, a type of star much larger than the Sun, called a red giant, has a low surface temperature of 3,500 K. The nearby bright blue star Bellatrix (on Orion's right shoulder) has a much higher surface temperature of 22,000 K. Bellatrix's light actually peaks in the UV but we see part of its light spilling over into the visible spectrum. Cool objects, like planets or asteroids, radiate at longer IR wavelengths.

Atmospheric filters

The blocking effect of the atmosphere to cosmic EM radiation is illustrated in Figure 10 by the curve of the height above the Earth's surface at which photons of a given energy get absorbed. The atmosphere is opaque to short wavelengths, as shown by the rise in the curve for photon energies more than a few eV. These are the γ-ray, X-ray, and UV photons, which have enough energy to ionize molecules in the atmosphere. Just beyond the blue end of the visible spectrum, the curve is particularly steep. This is because near-UV photons are absorbed strongly by ozone (O_3) molecules.

While the atmosphere is transparent to the wavelengths of visible light, the atmosphere is never completely dark. In the hours of darkness, the ion–electron pairs created earlier in the day by sunlight, recombine, and photons are emitted. The atmosphere also generates light when struck by energetic cosmic ray particles. This light is known as *airglow*, a pervasive dim light pollution that limits the faintest objects that ground-based telescopes can detect. Space telescopes, like the Hubble Space Telescope, are free from airglow and so can observe fainter objects.

Turning to longer wavelengths, atmospheric molecules such as water and carbon dioxide absorb most of the incoming (0.7–2 μm) infrared photons (a μm is a micron, or 10^{-6} m). These molecules have many vibrational spectral lines, and so give rise to the complicated opacity curve shown in Figure 10. For wavelengths longer than 5 mm, a second gap in the atmosphere opens up—a broad radio window—which extends to a wavelength of about ten metres. At still longer wavelengths, the frequency is low enough for the electrons in the ionosphere to screen out the wave fields, reflecting the waves back into space like a mirror. Radio operators exploit this effect when they communicate with stations beyond the horizon, by sending radio signals that bounce off the ionosphere.

Twinkling stars

The stars are so far away that when their light reaches the edge of the Earth's atmosphere, the wavefronts are almost perfectly flat. However, from there to the ground, the ride gets bumpier. The refractive index of the atmosphere is 1.0003, and so the propagation speed of light is a tiny bit less than c ($0.9997c$). Over a short distance, light refraction in air is barely noticeable, but after many kilometres of turbulent atmosphere it is significant.

Looking at a star through the atmosphere is like trying to peer through rippling water to see a penny on the bed of a stream. The atmosphere contains turbulent cells of high and low density, constantly fluctuating. *Atmospheric turbulence*—the bumpiness well known to air travellers—arises from the wind-driven mixing of warm and cool air streams. By the time starlight reaches the ground, its wavefronts are no longer plane—they are distorted, misshapen, and fluctuating. This is the reason stars twinkle. Another word for the effect is *scintillation.*

Optical astronomers quantify twinkling by the *seeing*—a measure of how much the atmosphere smears out point-like star images. The diameter of a stellar image is its 'seeing disc'; the smaller the disc the better the seeing. Typical values are around 1–2 arcseconds. (One arcsecond, denoted as 1″, is 1/60th of an arcminute, or 1/3600 of a degree.) The seeing varies with time, location, and with atmospheric conditions. At the best high-altitude astronomical sites, where the air flows smoothly, it can be as low as 0.5″.

The seeing values just quoted are at odds with a telescope's diffraction-limited angular resolution ($\theta = \lambda/D$), which we recall tells us that, at a certain wavelength, λ, the resolution can be made arbitrarily fine by increasing the diameter, D. For ground-based telescopes the seeing is limited by the atmosphere, and not by diffraction. Making a ground-based telescope larger than a certain size will bring no improvement to the angular resolution. With good seeing conditions, a simple telescope need be no larger than 20 cm, if all that is required of it is resolution. However, all is not lost. A powerful method for overcoming the seeing limit, called *adaptive optics*, is described in Chapter 6.

In Chapter 2, we took the eye as a model for a telescope and have used it to describe how light is collected, and how an image is formed using lenses or mirrors. But we have not yet talked about a vital part of the story—how light is detected. In the eye, detection is done by the retina. In telescopes operating in different parts of the EM spectrum, light interacts with matter in different ways, and so a variety of detectors is used.

Chapter 5
Instruments of light

Our eyes collect a continuous stream of photons from the world, and project images of it on to our retinas. One glance of the eye collects millions of picture elements, or *megapixels*, of detail, comparable with the number of sensors in a digital camera. A pixel is like one tile in a mosaic, a small part of a big picture. The surface of the retina is coated with photoreceptor cells. There is a 'sweet-spot' in the middle, the fovea, where they are packed in exceptionally tightly. There, the cells are around 2 μm apart, a spacing that matches the diffraction-limited resolution of the eye.

The eye functions something like a video camera that tends to track moving objects. Information from the photoreceptors is transmitted to the brain, via millions of nerves. Once an image arrives, it stays in the visual cortex for about 1/30th of a second, before the next one arrives. This is long enough to give us the impression that fast-moving objects move smoothly across the field of view.

If all the visual data flowing into the brain were stored at this rate, our 'wetware' processor, the brain, would be overwhelmed. Unlike the memory card in a camera, we don't store the contents of every pixel. Vision is much more about making sense of the world than about recording every scrap of visual data. Our brains cope with

12. A one-hour exposure of the Andromeda galaxy with its two dwarf companion galaxies, middle left and lower right, juxtaposed with a 0.01 second exposure of the Moon, to scale. The Moon is easily visible to the unaided eye, while, on dark nights, the much fainter Andromeda galaxy appears as no more than a fuzzy patch. (The author took these photographs with a 150 mm refractor and a CCD camera.)

the information torrent by inventing a model of the world to fill in the gaps. To do this, we have to make a priori assumptions about what we *believe* the world to be like, and subjective bias can enter. Often we don't actually see what we think we are seeing.
As soon as we move from a bright to a dark place, our eyes start to adapt to the new conditions. Our pupils dilate from 2 mm to around 7 or 8 mm—admitting over twelve times more photons. Over the next thirty minutes or so, chemical changes take place in the eye, resulting in virtually complete dark adaption.

Once adapted, and looking up into a clear dark sky, we can make out the faint sweep of the Milky Way. In the northern hemisphere we are aware of the hazy thumbprint of light from our nearest large galaxy neighbour, the great Andromeda spiral. In the

represented by a 2D pattern of electrical charge on the array. Each photosite has its own electrode, which, when activated, shifts the charge around, from one site to the next, along rows or columns, enabling the image to be read out, digitized, and stored.

Until only a decade ago, photographic plates were used to record images in wide-angle survey telescopes. For example, the 48-inch Oschin–Schmidt Telescope used 14-inch (36 cm) square plates, each containing, in effect, some half a billion pixels. However, as recording devices, photographic plates have now been largely superseded by mosaics containing many individual CCD arrays. The concept of tiling the image planes of telescopes with CCDs has led to the several-billion-pixel cameras planned for the next generation, such as in the *Large Synoptic Survey Telescope* (LSST) described in Chapter 9.

There are at least two ways to use CCD detectors. In a traditional 'point and shoot' imaging telescope, a target patch of sky is tracked and held centred in the field of view, while the camera's shutter is opened and closed. This system works well for both photographic and CCD cameras, but the quality of recorded images can be degraded if the telescope has tracking errors. The innate flexibility of CCDs has, however, provided an alternative approach. In a 'drift scan' mode, the telescope is stationary, while the image is allowed to drift across the focal plane. While the camera shutter is open, the data are continuously read out and stored. The CCD rows are aligned with the direction of the drift, and the charge image is stepped along the rows electronically to match the drift rate. Eventually the columns of charge reach the edge of the array and are read out—as strips of sky brightness. This is the method used in the *Sloan Digital Sky Survey* (SDSS) 2.5 m telescope located at Apache Point in New Mexico. Its camera has a mosaic of thirty CCDs, giving 120-megapixel images.

A pure silicon CCD functions from the near-UV to the near-IR, a wavelength range of 300–1000 nm. The high QE in this range means that a one-minute exposure with a CCD is equivalent to a one-hour exposure with photographic film. As with all detectors, there is unwanted noise, produced by thermal agitation of the silicon atoms. This noise is called the *dark current*, and can be reduced by cooling the wafer. By the late 1970s all major telescopes were using cooled CCD detectors for imaging, photometry, and spectroscopy.

Beyond the visible

Outside the visible range of wavelengths EM radiation interacts with matter in a number of different ways. At the longest radio wavelengths, the wave fields wiggle electrons in conductors (antennas), which we measure as oscillating currents and voltages. At the opposite end of the spectrum, the energies of γ-rays are so high that the photons punch their way through solids, leaving behind trails of debris as ionized matter. In this case detectors count the number of ion–electron pairs produced by the events.

The first indication that there was a form of light invisible to the human eye came in 1800, through one of William Herschel's greatest discoveries. He used a prism to disperse sunlight, and, with a thermometer placed beyond the red end of the spectrum, detected the presence of infrared radiation (IR) as heat. For the next century, the main method of detecting IR was through its heating effect. In 1822, Thomas Seebeck discovered that when two dissimilar metals are joined (forming a thermocouple), they generate a small voltage proportional to the temperature. To increase sensitivity, several thermocouples can be stacked, making a *thermopile*, a device that the English astronomer William Huggins used in 1869 to detect IR radiation from stars.

All hot bodies radiate infrared, and the Earth, the atmosphere, telescopes, and detectors are no exception. Given this contaminating thermal background, it is remarkable that Huggins was successful. What he achieved is akin to seeing stars in daylight. The trick was his invention of nodding a telescope, alternately looking at nearby blank patches of sky and subtracting the thermal background from that of the target. Modern ground-based far-IR and sub-mm telescopes use similar 'chopping and nodding' techniques. In these, a secondary telescope mirror is moved (chopped) at a few Hz (alternating between the target and an adjacent sky position). Then the whole telescope is nodded, on and off the target.

In 1880, Samuel Pierpoint Langley built a sensitive radiant heat detector known as a *bolometer*. The device relies on the increase of the electrical resistance of metals when heated, and consists of a thin, blackened metal foil. Langley's bolometer had a sensitivity of a ten thousandth of a degree, and could detect the heat from a cow a quarter of a mile away.

The low energies of IR photons make it difficult for them to be detected electronically. However, in the 1940s a detector was developed which, for the first time, responded directly to IR photons. The device was a spin-off from research into heat-seeking missile guidance systems, and is a thousand times more sensitive than a thermopile. Lead sulphide (PbS, or galena) is a semiconductor which, when cooled, has properties that make it sensitive to IR wavelengths as long as 4 μm.

Modern infrared astronomy began in the 1960s with a telescope having a mirror made from a spun 1.6-metre disc of epoxy resin which, when set, had assumed a parabolic shape and could be coated in aluminium. This was Gerry Neugebauer and Robert Leighton's telescope, equipped with PbS detectors, sited on Mount Wilson. By nodding the telescope at 20 Hz, they made a sky survey at 2.2 μm with a resolution of 2′. Some 20,000 IR-emitting stars and other bright sources (some with only faint optical counterparts) were found. The Orion star-forming nebula (M42) revealed an exceptionally bright source that is invisible optically—the *Becklin-Neugebauer object*. This is a stellar embryo—a *protostar*—hidden, cocooned inside its cloud of gas and dust.

Clouds of interstellar dust (stardust) pervade much of the Universe, obscuring the view at visible wavelengths. In galaxies the dust is often seen as dark lanes, silhouetted against diffuse background starlight (visible in Figure 12). The light from most of the stars in the Milky Way is absorbed and scattered by dust in the plane of the galactic disc, allowing only the nearest and least obscured stars to be seen.

When light is viewed through a cloud of particles, it is scattered and reddened. The process is called *Rayleigh scattering*, and involves short-wave (blue) light being scattered much more strongly than long-wave (red) light. When starlight is reflected from adjacent cosmic dust clouds this process produces blue-tinted *reflection nebulae*. An example is the wispy nebulosity surrounding the Pleiades star cluster (M45). On Earth there is a similar effect at sunset and sunrise. When the Sun is low in the sky, its light passes through a thick layer of atmosphere, reddening the sunlight, as the blue component is scattered away. But when the Sun is high, the scattered light colours the sky blue. One person's sunset is another's blue sky. The scattering effect becomes even weaker in the infrared region. This permits the interiors of cosmic dust clouds to be observed from the outside.

To detect IR, pure silicon can be doped with 'impurity' atoms (for example arsenic), rendering it more sensitive to low-energy photons. Cooled arrays of this material have been developed for use in the 5–29 μm wavelength range. Modern IR-sensitive devices include *photovoltaic* (PV) ones (in which IR photons generate electric currents), mercury-cadmium-tellurium (Hg-Cd-Te) alloys, and indium-antimonide (In-Sb) diodes.

Objects or regions which are cooler than 100 K, like certain galaxies, protostars, and the protoplanetary discs of debris around stars, emit at far-IR and sub-mm wavelengths (15 μm–1 mm). Gallium-doped germanium (Ga-Ge) bolometers have been developed for this very interesting part of the spectrum. When cooled to nearly absolute zero, the detectors become sensitive to wavelengths as long as 115 μm. At very low temperatures, some materials lose all electrical resistance and become superconductors. They can be made to 'flip' in and out of the superconducting state by a tiny change in temperature, and so can be made extremely sensitive to the IR heat flux. These are *transition edge superconducting* (TES) bolometers.

Detecting high-energy photons

At short wavelengths, there are three basic interactions between photons and matter. The first is *photoelectric absorption*, which, as we saw in Chapter 4, occurs in the upper atmosphere. It involves the absorption of a photon by an atom resulting in electron excitation to a higher energy level, or its ejection to form an ion–electron pair.

The second is *Compton scattering*. In 1923, the American physicist Arthur Holly Compton found that when energetic X-ray photons collide with stationary electrons, they lose energy, and emerge with longer wavelengths. This showed that photons and electrons exchange energy and momentum, as if they were particles, and confirmed Einstein's quantum picture of light.

The third process is *pair production*, which occurs for the highest energy γ-ray photons, above 10 MeV. In this, a photon passes close to an atomic nucleus and interacts with the nuclear electric field to produce two particles, an electron and a *positron*. The positron is *antimatter*. (An anti-electron is a positively charged electron.)

The simplest type of X- or γ-ray detector is an *ionization chamber*. When a high-energy photon ploughs through matter, it can produce many ion–electron pairs. This collisional debris can be detected as an electrical current, using a gas-filled container with a voltage applied to electrodes. A *Geiger–Müller* tube is similar, except that, with a higher voltage, it triggers an avalanche of particles, producing a stronger signal. Another variant, the *proportional counter*, is operated at an intermediate voltage, allowing the number of ion–electron pairs to be counted and so measure photon energies.

CCDs can also detect X-ray and UV photons. At visible wavelengths, one photon produces one photoelectron. But a higher energy photon can also produce secondary electrons in the device. A CCD detector can therefore provide simultaneous imaging and energy measurement, where the charge stored at each photosite is proportional to the photon energy. The *Chandra* X-ray space observatory, for example, carries CCD array detectors to image X-rays with energies between 0.2 and 10 keV. This enables the X-ray emission from different chemical elements such as oxygen or iron to be separated.

Compton scattering and photoelectric absorption are important processes for energetic photons (over 100 keV). Such photons can be detected when they 'Compton scatter' from electrons in a scintillator detector. In a solid *scintillation counter*, an energetic photon strikes a crystal, producing a microsecond-long light flash, which can be detected with a photomultiplier.

At very high photon energies, it is possible to track particle trajectories in a *spark chamber*. In this device, γ-rays pass through a gas-filled chamber containing an array of electrically charged metal plates. When a pair-production event takes place, the particle tracks show up as lines of sparks which can be 'back-projected', pointing to the source of the γ-ray.

From the rainbow to the stars

Each chemical element has a unique identifying spectral signature, a fingerprint. An atom in a low-energy state can be excited to a higher one by absorbing a photon with a precise frequency. The opposite occurs when an electron in an excited state makes a transition to one of a lower energy, by emitting a photon. It is easy to demonstrate spectral line emission, by throwing a pinch of household salt on a flame—it will flare up bright yellow. The characteristic yellow (the same colour as in sodium streetlights) comes from the specific wavelength emitted when electrons in sodium atoms jump between energy levels.

Flame spectra of chemical elements and compounds were studied in the early 19th century, when chemistry was in its infancy. In these experiments, glass prisms, similar to those used by Newton, were used to split or disperse light into its colours. In 1665, Newton had made a hole in a window shutter to let in a beam of sunlight (the brightest light source then available), to make a spectrum. With a screen and aperture, he isolated a single colour, and passed the light through a second prism. The light could not be split any further, and the experiment showed that white light is a mixture of all the colours of the rainbow.

In 1802 an English chemist, William Wollaston, observed the solar spectrum, and found some narrow dark lines slicing into it, a discovery that went almost unnoticed at the time. A decade later, Fraunhofer developed a spectroscope with a high spectral resolution, by using a small telescope to focus the light on a slit in front of the prism. Turning it towards the Sun, he discovered several hundred dark *Fraunhofer lines*, superimposed on the colours of the much smoother underlying spectrum. Gustav Kirchhoff and Robert Bunsen noticed that some of these dark lines had the same wavelengths as bright spectral lines, seen with chemical elements in laboratory flame spectra.

When a continuous spectrum, like that from the Sun, is seen through a layer of cool atoms, these can absorb photons with specific energies, but only those corresponding to the energies of quantum atomic transitions of the atoms present. It became clear that the narrow, dark Fraunhofer lines were produced by the *absorption* of specific wavelengths of the continuous spectrum by particular atoms in the solar atmosphere, and this allowed the chemical elements present to be identified.

By 1870, Kirchhoff had used spectroscopy to identify many chemical elements in the Sun. However, in 1868 the English astronomer, Norman Lockyer, observed a line in the solar spectrum that had no known earthly counterpart. He named it *helium* (from the Greek for Sun). Helium, the lightest noble gas, was discovered on Earth some forty years later.

Although glass prisms are relatively easy to make, they cannot match the spectral resolving power and efficiency of a diffraction grating. An ordinary compact disc behaves like a diffraction grating, reflecting and separating the colours of white light. The light is scattered by the lines of the tracks, which are spaced 1.6 μm apart. In a diffraction grating, many fine parallel lines are ruled on the surface of a flat plate. Light can either be reflected (in a reflection grating, like the CD) or transmitted through it (as shown in Figure 13).

In Figure 13(a), plane wavefronts of blue light illuminate the slits of a grating. The light is diffracted through the slits, emerging as

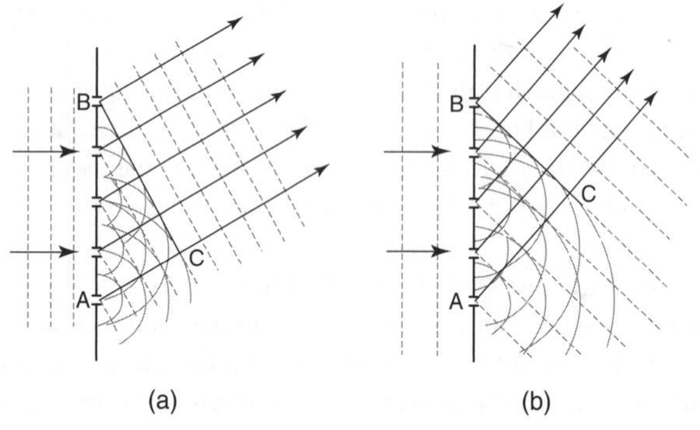

13. A transmission diffraction grating illuminated with (a) blue, and (b) red light. Some of the Huygens wavelets are shown.

spherical Huygens wavelets. These interfere strongly in certain preferred directions, for example along the wavefront BC. In these directions, the path differences, such as AC, must be a whole number of wavelengths. But the reinforcement in that direction occurs only for blue light. For longer wavelength red light (Figure 13(b)), the reinforcement direction is at a different angle; each wavelength of light is scattered strongly into its own angle.

The spectral resolving power of a diffraction grating increases with the number of lines it has. A typical high-resolution grating can separate two wavelengths differing by as little as one part in 100,000, which is roughly ten times narrower than is possible with a glass prism.

By the end of the 19th century, telescopes equipped with spectrographs were routinely used to study the stars. Normal stars like the Sun contain about 72 per cent hydrogen, 25 per cent helium, and 3 per cent metals. When astronomers refer to 'metals' they mean elements heavier than hydrogen and helium. The distinction is that hydrogen and helium are primordial (made in the Big Bang), and the other elements were (and continue to be) synthesized in stars.

Spectroscopy can also be applied to study galaxies. Different galaxy types contain different stellar populations, which show up like fingerprints in their spectra. Some galaxies are actively forming new stars, whose blue and ultraviolet light dominates spectra. Other galaxies have ceased forming new stars and their stellar populations are old, emitting much redder light, often extending into the infrared. Galaxy spectra can show the absorption lines of chemical elements in the interstellar medium, ejected from dying stars.

Spectral measurements can reveal the radial velocity of a galaxy (its redshift or blueshift), as well as the internal motions of its stars and gas. The mass of a rotating disc galaxy, like the Milky Way, can be inferred from these measurements by balancing centrifugal and gravitational forces. An instrument that can simultaneously collect spatial and spectral data is a *long-slit spectrograph*. As the name suggests, the instrument has a long, narrow entrance slit, which is positioned in the focal plane of a telescope. Light from different parts of the galaxy is dispersed into a series of spectra, ranged along one spatial dimension.

For measuring large extended objects, a spectrograph slit can, in principle, be stepped across the sky to sample spectra from a 2D region. However, in practice this is prohibitively time consuming. A more efficient instrument is an *integral field spectrograph* (IFS), combining imaging in 2D with spectroscopy. One type of IFS is based on placing a 2D array of tiny lenses (*lenslets*) in the image plane of a telescope. These represent the pixels. The light from each pixel is separately dispersed and imaged, giving each pixel its own spectrum.

In Chapter 5 we have seen how, first, photographic, and then electronic light detectors have greatly increased the sensitivity of telescopes, and made it possible to record the state of the sky, as well as obtain information on the physical properties of stars and galaxies. Telescopes and their detectors are now inseparable. In Chapter 6 we will begin to assemble these pieces, and examine how the convergence of several diverse technologies has, in the last few decades, produced the *very large telescopes* (VLTs).

been digitized and extracted to computer databases using plate-scanning machines. Legacy photographic surveys like this contain astrometric, colour, and galaxy morphology data, but generally there is no redshift information.

With modern synoptic telescopes, advanced detectors, like the IFS described in Chapter 5, can be incorporated to obtain spectral and spatial information. This of course means more data. For example, a 15-megapixel CCD camera can produce a 16-bit monochrome image of size 30 MB, with 160 images stored on a DVD. If spectral information is added, for example, by dispersing each pixel's light into 10,000 wavelength intervals, a *data cube* is produced; in this example it would contain 300 GB. The data structure can be thought of as a stack of 10,000 2D images, each taken at a different wavelength. Some 64 DVDs would be needed to store a cube like this.

When both spatial *and* redshift data are available, galaxies can be mapped to their 3D positions in space using redshift data as a proxy for distance (via the Hubble law). An example of the power of this approach is from the 4-metre Anglo-Australian Telescope, which has a 2°-wide-field robotic spectrograph, used to produce the *two-degree field (2dF) surveys* and measure the spectra of 250,000 southern hemisphere galaxies. The results show galaxies are distributed very non-uniformly; they are clumped in large aggregations and sheets, separated by enormous empty voids—like cosmic foam.

In measuring galaxy redshifts, spectrographs certainly give highly accurate results, but for large surveys, analysing the spectra of many pixels can be time consuming. Also, when the spectral resolution is high, the sensitivity per wavelength interval is smaller and only the brighter objects can be measured.

However, there is a much quicker method, that of determining *photometric redshifts*. In their rest frames of motion, the spectra of galaxies contain characteristic features, or 'breaks' in luminosity. The redshift of a galaxy alters the wavelengths at which these features are observed. By measuring the apparent brightness of the galaxy with a set of different coloured filters, and fitting model spectra to the data, the redshift can be estimated. The filters behave like a coarse diffraction grating and, although the resulting errors are larger than for spectroscopic measurements, photometric redshifts can be used for extremely faint objects.

The SDSS telescope began surveying a quarter of the sky in 2000, and has now amassed data on half a billion galaxies with its 1.5° field of view. One of its cameras uses five standard optical filters to measure redshifts photometrically. The data produced by the SDSS exist in an open-source repository. The classification of the morphology of SDSS galaxies was one of the first citizen science *Galaxy Zoo* (*Zooniverse*) projects, set up in 2007. The future role of crowdsourcing astronomical data will be discussed in Chapter 9.

The role of smaller telescopes

Everything described so far has been about the headlong push towards ever-larger telescopes. While it is true that VLTs enable us to see the most distant objects, there remain important roles for smaller (sub 2 m) telescopes. These include observing 'targets of opportunity' and following-up transient objects, such as supernovae, γ-ray burst (GRB) afterglows, and near-Earth objects. Small telescopes are often operated remotely. A small robotic telescope, like the 2-metre *Liverpool Telescope* (on La Palma in Spain), can be pointed rapidly at a transient source. If a source is fading fast, a small nimble telescope like this can collect more photons than a larger, slower one.

Small telescopes can also discover exoplanets. One technique is the *radial velocity* (or *Doppler spectroscopy*) method, in which high-resolution radial velocity measurements are made of a star. When a planet orbits the star, the two bodies revolve around the

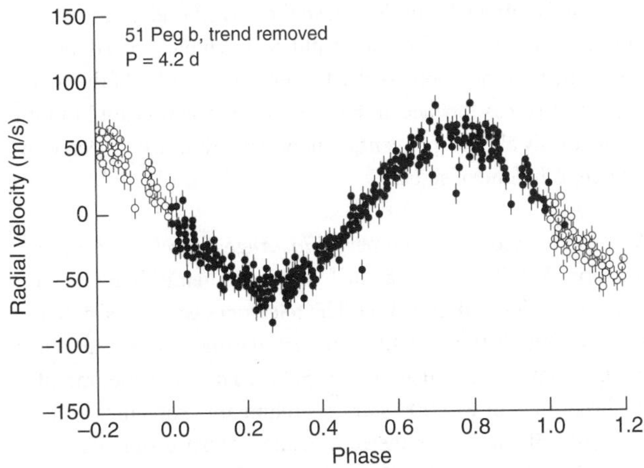

17. The radial velocity of the star 51 Pegasi b, showing the periodic trace indicating the presence of an exoplanet.

common centre of mass. From a distant point in space, the star appears to wobble periodically, resulting in a regular Doppler shift heartbeat of its spectral lines. The first exoplanet orbiting a normal star (51 Pegasi b) was discovered in 1995 (Figure 17) using a high-resolution spectrograph on the 1.9-metre telescope at the *Observatoire de Haute Provence* in France. The exoplanet was the first *hot Jupiter* to be found—a large planet orbiting close to its parent star.

A large planet which lies close to its star produces a large wobble, but the wobble gets weaker and harder to detect for the interesting lower mass, and potentially habitable Earth-sized planets orbiting further away from their stars.

Another method, *planetary transits*, relies on the tiny dimming of a star's light when a planet passes in front of it. It is relatively easy to detect large planets this way. For example, viewed from a distant point in space, a transit of the giant planet Jupiter dims the Sun's light by 1 per cent. The diameter of the Earth is only about one tenth of Jupiter's, and so the corresponding transit dimming is 0.01 per cent and would be much harder to detect. The transit method only works, however, if the orbit of the planet happens to cross the face of the star. This is a significant limitation because only about 1 per cent of planetary orbits are expected to intercept the line of sight.

A pair of telescopes, the (super) *Wide-Angle Search for Planets* (or super-WASP), covering north and south hemispheres, is based in South Africa and La Palma. The telescopes each consist of a stack of eight 200 mm, f/1.8 commercial camera lenses, attached to CCD cameras. The field of view is 7.8°. The telescopes monitor the brightness of 100,000 stars, searching for exoplanets. Super-WASP has, so far, discovered about thirty exoplanets, a high scientific return for a low investment.

Global networks of small telescopes can also provide a flexible system for observing transient phenomena. For example, the *Las Cumbres Observatory Global Telescope Network* is setting up forty robotic 0.4–2-metre telescopes, to provide rapid sky coverage at all latitudes and in both hemispheres. The activity will include targeting transient sources as well as searching for exoplanets by the radial velocity method.

We have been looking at some of today's remarkable optical and IR large telescopes, highlighting the technologies that made them possible and the science results that have revealed galaxies at vast distances and, closer to home, exoplanets. In Chapter 7, we will keep our feet on the ground, but with telescopes that can see wavelengths a million times longer—radio telescopes.

Chapter 7
The radio sky

In 1932, Karl Jansky discovered cosmic radio waves, quite by accident. Jansky was a radio engineer working for Bell Telephone Laboratories in New Jersey, trying to locate a pervasive and troublesome interference (radio static), which had been plaguing a transatlantic radiotelephone service, at a wavelength of fifteen metres. The waveband is frequently noisy, and at the mercy of the changing state of the ionosphere. To try to pinpoint the noise, Jansky built a large rotatable antenna. When Jansky rotated the antenna, he found that the static peaked in one direction. Fortunately, that year there was a minimum in solar activity, which made it much easier him to recognize the origin of the noise. By careful measurement, he discovered that the peak of the noise shifted direction by about one degree a day—the sidereal rate. The source lay outside the solar system, towards the centre of the Milky Way, in the constellation of Sagittarius.

The first radio telescope

An amateur radio enthusiast, Grote Reber, read about Jansky's discovery and, in 1937, built a 9-metre paraboloidal reflecting dish, from timber and metal sheets in his Illinois back garden. This was the first radio telescope. The dish could be tilted in elevation, and relied on the Earth's rotation to make drift scans of the sky. At the focus of the dish, the radio energy was picked up by antennas connected to radio receivers at wavelengths of 180 and 63 cm. Reber recorded the telescope's signal on a paper strip chart recorder, and made maps of the 'cosmic static', which he published between 1940 and 1948.

During World War II, technology for radar systems leaped forward—from metre to centimetre wavelengths. In 1942, James Hey, one of the early British radar pioneers, was looking for strong signals, thought to be jamming long-wavelength radars in the English Channel. He discovered that the signals came from *solar flares*. These are violently energetic releases of charged particles and magnetic fields from sunspots, producing intense broadband radio emission.

After the war, the scientists who developed radar, including Bernard Lovell and Martin Ryle, made use of the advances in electronics, radio technology, and digital computers, to found a new science—radio astronomy. Single-dish antennas continue to play important roles. One of the most famous is the fully steerable *Lovell Telescope* (Figure 18) built in 1957 by Bernard Lovell at Jodrell Bank Observatory near Manchester, England. It can now operate at wavelengths down to 6 cm, where it has an angular resolution of 3′.

An even larger 1,000-foot (305 m) spherically shaped dish is at the *Arecibo Observatory*, in Puerto Rico. It was built in 1963 in a karst sinkhole, a natural geological depression. To follow sky objects, a moveable tracker is suspended 150 metres above the dish, containing a simple two-element spherical aberration corrector and antennas. As the tracker moves, it picks up radio waves from different parts of the dish. (This is the paradigm on which the HET and SALT telescopes are based.) Arecibo operates at wavelengths 3–100 cm, and, with its large collecting area and high sensitivity, is well suited to observing transient sources within a 40° strip of sky overhead.

18. The 250-foot dish of the Lovell Radio Telescope.

A problem with a single-dish radio telescope is that its angular resolution is intrinsically poor, a consequence of the long radio wavelengths and the difficulty in making large dishes. The beam width of Reber's telescope was about 13°, at a wavelength of 180 cm, a very blunt tool compared with the Hale Telescope. From Chapter 2, recall how a telescope's diameter, expressed in wavelengths, determines its diffraction-limited angular resolution. For the 200-inch telescope the number is ten million; for the Lovell Telescope it is a thousand. The question is: how could a radio telescope bridge such a huge gap to make observations with high angular resolution?

Interferometers

The answer is to operate two antennas as an *interferometer*. An interferometer has an angular resolution limited by the spacing D between the antennas, and not by the size of the individual antennas. The technique of interferometry, at optical wavelengths, has been known about since the 19th century. In 1920, Albert Michelson and Francis Pease built a successful optical interferometer with two 15 cm mirrors, on a 6-metre-long strut attached to the Hooker Telescope to measure the diameter of stars. With this, they measured the diameter of Betelgeuse as 0.047″.

The operation of an interferometer for radio wavelengths is shown in Figure 19(a). The incoming wavefronts (dashed lines) lie in the plane BC. A wavefront arrives first at antenna B, where it produces an oscillating current. After travelling an extra distance AC (the path difference) the wavefront arrives at the second antenna A. When the signals at A and B are added, the result depends on the path difference AC, which determines the relative phase of the signals. If AC is a whole number of wavelengths, the two signals add coherently and the combined amplitude is doubled; if it is an integral number of half wavelengths, the combined amplitude is zero.

Take the Sun for example. It emits radio waves, both from its sunspots (which are small) and from its corona, which is much larger. As the Earth rotates, these sources sweep across the sky. If a radio interferometer is arranged along an east–west line, the angle of the arriving wavefronts (and path difference AC) varies in time. In the case of a sunspot, the output of the interferometer traces a modulated pattern of constructive and destructive interference fringes—peaks and troughs (shown inset in Figure 19a). The fringes follow the time-varying path difference, as the sunspot's wavefronts come in and out of phase at the two antennas. The *visibility* of the sunspot fringes—the relative difference between peaks and troughs—is large, indicating that the sunspot is only partially *resolved* by an interferometer with this spacing. (The fringes produced by a radio interferometer are analogous to those in Young's two-slit experiment, see Figure 3(a).)

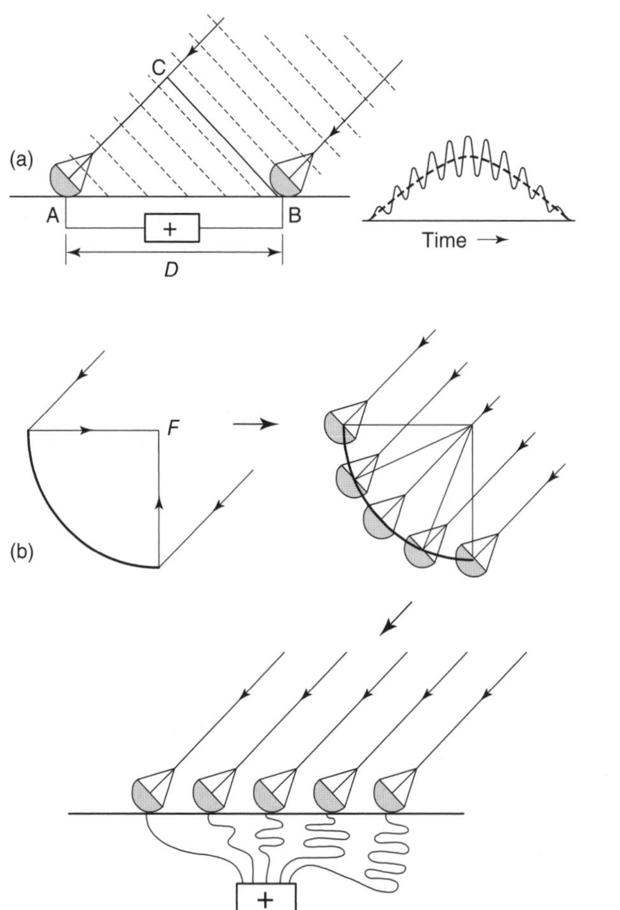

19. (a) A two-antenna interferometer with baseline *D*. The signals are connected via cables to a receiver, where they are added (+), producing the time-trace shown inset; (b) synthesizing the aperture of a giant dish of focal length *F*. (Right) dividing the dish's surface into a number of separate smaller equivalent antennas. (Lower) the small antennas are now placed at ground level, connected to a receiver with different length cables, simulating the time delays of different parts of the big dish.

But when the source is much larger, as it is for the solar corona, the interferometer behaves differently. The increased range of angles from the antennas to the corona means that, wherever the Sun is in the sky, there will always be *some* waves from its corona arriving in phase at both antennas, which add constructively. The time-trace of the receiver is much smoother (indicated by the dashed line inset in Figure 19(a)) and the visibility of the fringes much lower. An interferometer with this spacing is said to have *resolved* the source.

At this point, it is worth referring back to the two sinusoidal waves of Figure 3, and to think about how waves can be added. There are clearly many ways in which this can be done, with different wave amplitudes and phases. In 1822, the French mathematician, Joseph Fourier, proposed that any mathematical curve or function could be represented, by adding up many waves (these are known as Fourier components). The ones shown in Figure 3 are two very elementary examples of Fourier's powerful method; the waves on the left are the Fourier components, and they add up to produce the ones on the right.

To describe complex curves, more than two Fourier components are needed. The sound of a bowed violin string (a pressure wave in the air) can, for example, be represented by the sum of many Fourier components. Fourier's method is not limited to the time and frequency domains, or just to one dimension. A 2D example is an image. Any image can be represented as the sum of its *spatial* Fourier components (that is, by a set of periodic waves of intensity spread over an image plane).

What has this got to do with how an interferometer works? When the instrument is used to measure a radio source and produce fringes it is, in effect, measuring just one of the spatial Fourier components of the sky brightness distribution. The angular width of the component in question is given by the diffraction formula: $\theta = \lambda/D$.

Returning to the example of the sunspot, the high fringe visibility tells us that there is a small, unresolved source present, with an angular size less than $\theta = \lambda/D$. By gradually increasing D, and observing the way the fringe visibility changes, we can find the interferometer spacing where the visibility begins to fall and the source starts to be resolved. This tells us how big the sunspot is. Conversely, the low fringe visibility of the Sun's corona is telling us that this source has been resolved. If we want to measure the size of the corona, we would need to make D smaller, and find the point at which the fringe visibility starts to increase.

Martin Ryle and Derek Vonberg first demonstrated these features with a radio interferometer in 1946, in Cambridge, England, and determined the size of sunspots. They also measured the absolute brightness of the emission, and found it to be very much larger than the blackbody level expected from the surface of the Sun. They concluded that the emission must have a non-thermal origin.

From this time onwards, radio telescopes revealed many bright, unresolved sources in the sky. The sources produced strong interference fringes, even when the antennas were widely separated, showing that some sources were extremely small. For a while, the sources were called 'radio stars'. However, uncertainties in their positions made it difficult to identify them with any optical objects. As we will see shortly, these were not signals from ordinary stars.

There is a simple way to measure the positions of some radio sources accurately. When the Moon passes in front of a source, the source is occulted, and the radio waves are (temporarily) blocked. The timing of the occultation can be used to give an accurate source position, since the position of the Moon's limb is well known. In 1962, Cyril Hazard succeeded in pinpointing the bright radio source 3C273 by *lunar occultation* using the 64-metre dish of the *Parkes radio telescope* in New South Wales, Australia. The position of 3C273 was sufficiently accurate for Maarten Schmidt to identify the optical counterpart using the 200-inch Hale Telescope and measure its redshift (as $z = 0.16$). This was the first quasi-stellar object, or *quasar*, to be discovered. The high redshift of the galaxy hosting the radio source implied a (then unheard-of) distance of 2.5 billion light years. The discovery of such a bright object, so far away, also implied a vast energy source—four trillion times the luminosity of the Sun. This was no ordinary star.

We now know that a quasar is a small emitting region in the centre of a galaxy, an *active galactic nucleus* (AGN). An enormous amount of energy is released when matter falls on to a central supermassive black hole (SMBH). Most galaxies are now believed to contain SMBHs in their nuclei, including the Milky Way. Some of them, like 3C273, are radio-loud quasars, and some of them are radio-quiet. These differences can be understood if the powerhouse (the central SMBH) is surrounded by a thick dusty torus. Different galaxies are viewed from a variety of angles; from some directions the nucleus is obscured by dust; from others it is exposed.

Aperture synthesis

Small dishes, paired as interferometers, give information about the angular size of sources in the direction along the baseline of each interferometer. It is possible to combine data from a number of different baselines and angles to make images of radio sources with a resolving power equivalent to that of a giant single dish, with overall diameter D: *aperture synthesis*. In this case D refers to the diameter of the whole array. Each baseline contributes one spatial Fourier component to the final image; these data are added, in a digital computer, to build the final 2D image, or map of the sky brightness. The angular resolution of the 2D map is $\theta = \lambda/D$.

Aperture synthesis is illustrated in Figure 19(b). An imagined giant dish focuses an incoming wavefront to the focus F, and we divide up the surface of the dish into smaller areas, where small antennas are placed. The signals from these are sent to a receiver, to be summed. Provided the incoming wavefront signals are added with the correct phase (using connecting cables of appropriate lengths to adjust for the time delays), the process is equivalent to that of a big dish. Any pair of small antennas is a fringe-producing interferometer, capturing one Fourier component of the sky brightness. In the lower diagram the antennas are shown moved to ground level, with cable lengths adjusted to replicate the relative time delays of the ray paths in the large dish (a phased array).

There is no need to cover the *whole* area of the synthesized dish with small antennas, nor is it necessary to acquire all the signals simultaneously (provided the source is not varying faster than the observation timescale). Instead, parts of the dish can be sampled, and a smaller number of antennas moved around to build up the radio image gradually.

Moving large numbers of antennas around is not very practical. One way to speed things up is to arrange dishes along an east–west line, and let the Earth's rotation do some of the work: *Earth rotation synthesis*. This provides relative motion of the antennas (as seen from the fixed stars). The process can then be repeated, with different spacings, until all parts of the big dish have been sampled. Martin Ryle pioneered the technique at the *Mullard Radio Astronomy Observatory* (MRAO) in Cambridge, England, using three steerable dishes (the *One-Mile Telescope*) riding on a section of the defunct Oxford to Cambridge railway line. An early digital computer (Edsac II) was used to convert the interferometer data into the first aperture synthesis radio map, the *North Pole Survey* in 1962. This feat was no less significant than the one achieved, about a century earlier, when the first photograph of the heavens was taken through a telescope.

Even longer antenna baselines, and shorter wavelengths, were then explored. In 1976, a *Multi-Element Remote-Linked Interferometer* (MERLIN) was constructed, consisting of seven large radio dishes spread 217 km across England, linked to Jodrell Bank. Initially, microwave links were used to connect the antennas to a central receiver, but, in 2009, optical fibre cables replaced the links (e-MERLIN), giving a hundred-fold increase in bandwidth and higher sensitivity. The telescope now operates between 1.4 and 22 GHz, giving an angular resolution of about 0.04″, equal to that of a VLT using adaptive optics.

The *Karl G. Jansky Very Large Array* (JVLA, 1980) telescope in Socorro, New Mexico (Figure 20), is a large-aperture synthesis telescope consisting of twenty-eight 25-metre dishes, laid out as a 'Y' with 21-km long arms. The JVLA antennas can be moved to various locations on railway tracks allowing apertures up to 36 km to be synthesized. The telescope operates at wavelengths from 6 mm to 4 metres. At the shortest wavelength, the angular resolution is 0.05″. In 2008, the bandwidth of the telescope was increased (and the sensitivity, by a factor of ten). The JVLA has remained in the front line of radio astronomy for over three decades.

20. The Karl G. Jansky Very Large Array (JVLA) radio telescope in New Mexico.

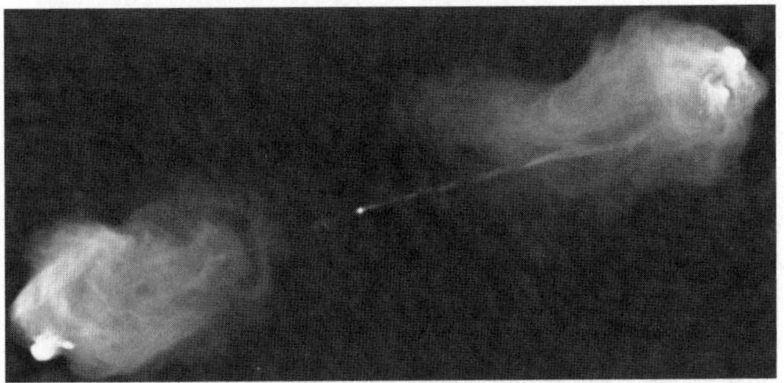

21. The very large array 6 cm map of one of the brightest radio sources: the radio galaxy Cygnus A (3C405). The nebulous outer radio-emitting twin-lobes are produced by synchrotron radiation, as highly energetic particles in the narrow jets plough into the external medium. The small 'point' in the centre coincides with the galaxy containing a supermassive black hole powering the emission.

Opening up a new radio window on the Universe revealed exotic sources. For example, many radio galaxies show characteristic double-lobed and jet-like structures. A well-known example is Cygnus A, 600 million light years away (Figure 21). Radio galaxies like this show two key features: a power law spectrum (the brightness decreases with frequency), and polarized radiation. These features are inconsistent with blackbody emission. A non-thermal process, *synchrotron radiation*, can instead explain them. In a typical non-thermal process, a special group of unusually energetic particles is accelerated, generating EM radiation of exceptional intensity. In a radio galaxy like Cygnus A, electrons, moving near the speed of light, gyrate in cosmic magnetic fields and emit polarized radiation.

Very long baseline interferometry

To achieve the very highest angular resolution, radio interferometers are operated with baselines of hundreds or thousands of kilometres; the technique of *very long baseline interferometry* (VLBI). Over these large distances, the relative signal phase is preserved by using radio links, fibre optic cables, live streaming over the Internet, or by digitally recording signals at each antenna with accurate clock signals.

Using short wavelengths (around 1 mm), the *Event Horizon Telescope* is an Earth-sized telescope, consisting of several dishes located around the world. These form a precisely phase-matched interferometer capable of resolving features as small as 15 microarcseconds. This is equivalent to seeing an object the thickness of a human hair in Rome, from London. This resolution is thought to be enough to see the event horizon of the SMBH Sagittarius A* at the centre of the Milky Way, and image it against the background of the hot gas surrounding it.

To extend baselines further, antennas have even been carried on satellites, with their signals correlated with those on the ground. This was first done in 1997 with the Japanese *VLBI Space Observatory Programme* (VSOP). At one end of the 21,000 km interferometer was an 8-metre space antenna, giving a resolution of less than a milliarcsecond. In 2011, a Russian 10-metre antenna was carried on the *Spectr-R* satellite, achieving baselines of up to 200,000 km, and a resolution of a microarcsecond.

Pulsars

In Chapter 3, we saw how atmospheric turbulence makes stars twinkle. Bright quasars also scintillate at radio wavelengths—but for a different reason. In reaching our telescopes, radio waves pass through the *solar wind*—a stream of turbulent plasma flowing out of the Sun. The solar wind has an irregular, lumpy texture and radio waves passing through it are refracted, producing scintillation. This was first observed in 1951 by the English radio astronomer Anthony Hewish. To study the solar wind, Hewish and his graduate student, Jocelyn Bell, designed and built a low-frequency (81 MHz) *four-acre telescope*, consisting of a phased array of over 2,000 wire dipoles (and reflecting screen) at the

MRAO. The telescope had the fast time response (0.1 second) and large aperture needed to detect rapidly varying signals. Observations were started in 1967 and Bell quickly discovered an unusual source that was pulsing regularly, about once a second. The pulses appeared to be coming from a fixed direction in the sky. This was the signature of the first pulsar, and three more pulsars, with different periods, were found soon after. Hewish and Ryle were jointly awarded the 1974 Nobel Prize, for the discovery of pulsars and the development of aperture synthesis, respectively.

It became clear that the pulsar's regular beat was coming from a rotating object—like flashes of a lighthouse beam. But the pulses arrived too rapidly to have come from a normal star; the object must be tiny. But why wasn't it ripped apart by centrifugal forces? If gravity were holding it together, Bell's object would need to have mass density much higher than any normal star. The most likely candidate was a *neutron star*, an object that had been predicted theoretically in 1934 by Walter Baade and Fritz Zwicky, but had never actually been seen.

A neutron star is formed when a massive star explodes—a supernova. At the end of the star's life, when the fuel has been used up, the pressure supporting it against gravity suddenly disappears and the core of the star collapses dramatically—in seconds. The collapse releases an immense amount of energy, explosively ejecting the star's outer layers and leaving behind a spinning compact object, the neutron star. Made almost completely from neutrons, a neutron star has a density similar to that of the atomic nucleus. A teaspoonful of neutron star would weigh 5 billion tonnes.

When the star's core collapses, the magnetic field threading through it is compressed by an enormous factor. The spinning, vastly amplified magnetic field generates a huge electric field at the surface, accelerating electrons up to nearly the speed of light. The electrons gyrate in the magnetic field, producing non-thermal

EM radiation. A neutron star has a mass of at least 1.4 times the mass of the Sun (300,000 times the mass of the Earth), but its tiny radius—about 10 km—is the size of an average city.

In 1974, American astronomers Russell Hulse and Joe Taylor discovered PSR B1913 + 16, the first *binary pulsar*, using the Arecibo telescope. They noticed that the pulse rate varied regularly with a period of 7.75 hours (with a faster rate when the pulsar is approaching us, and a slower one when receding, in line with the Doppler shift). They realized that the pulsar had a companion, another neutron star, and that the two are orbiting around their common centre of mass at high velocities. The very strong gravity in the binary pulsar system is ideal for testing Einstein's 1915 *general theory of relativity*. Here, there is a precise clock (the pulsar) orbiting in a strong gravitational field of the neutron star. The masses of both stars could be estimated with an exquisite precision of less than 0.05 per cent—only equalled in astronomy by our knowledge of the mass of the Sun.

When collapsed massive bodies (like these orbiting neutron stars) move, they warp space-time, creating ripples that spread out at the speed of light. These ripples are *gravitational waves* and are predicted by Einstein's theory. Theoretically, the binary pulsar should radiate gravitational wave energy, steadily slowing the orbital period down by a small, measurable amount. The Arecibo telescope has monitored the binary system since 1974, and found that the orbit of PSR B1913 + 16 is behaving exactly as theory predicts. Hulse and Taylor were awarded the 1993 Nobel Prize for this discovery.

Although gravitational waves had not actually been detected in 1974, on 14 September 2015 the two detectors of the Laser Interferometer Gravitational-Wave Observatory (LIGO), based in Hanford, Washington and Livingston, Louisiana, almost simultaneously detected a transient gravitational wave signal (from an event known as GW150914). The signal has been

Jupiter—as well as find a large previously unknown ring around Saturn.

The largest IR telescope so far launched (with a 3.5-metre mirror) was ESA's *Herschel Space Observatory*, operating from 2009–13 from its L2 point. It covered a wide wavelength range: 50–500 μm. Herschel's missions included making the largest ever survey of cosmic dust, determining the chemical compositions of solar system objects, and identifying molecules in galaxies. Herschel has found a large population of galaxies radiating at sub-mm wavelengths, which do not fit current models of star formation in galaxies. In 2013, it discovered an ancient and unexpected starburst galaxy (HFLS3), 13 billion light years away, making stars 2,000 times faster than the Milky Way is currently doing.

The cold Universe: the cosmic microwave background

Some of the most important discoveries in astronomy have been made by chance. Looking with the right telescope, at the right time, and at the right wavelength is only part of the story. When faced with a result that does not fit into the accepted scheme of things, the discoverer (or discoverers) have, above all, to have the sheer determination to eliminate all false positives, and pursue a genuine result to the bitter end. This was the certainly the case with *cosmic microwave background* (CMB) radiation.

The discovery of the recession of the galaxies in the 1920s, embodied in the Hubble law, seemed to imply that the Universe had once existed in a hot compact phase—the Big Bang, as it later came to be called. But there was, at the time, no direct proof that this had actually happened. If there had been a hot phase, it was argued, the Universe should now be filled with primordial thermal radiation, a remnant of the earlier fireball. In the ensuing expansion, the thermal radiation would now have a greatly reduced temperature, of a few degrees K, and its photons would be the oldest ones in the Universe.

In 1960, NASA launched a metallized communications balloon, *Echo I*, as a passive microwave reflector, to test long-distance communications. Signals were bounced off the balloon, to a remote station. The signals were so weak that a sensitive receiver and antenna was needed. For this, the *Holmdel 6-m Horn Antenna* was constructed at Bell Telephone Laboratories in New Jersey. The antenna resembled a huge steerable ear trumpet, to collect and swallow the incoming radiation down its throat, to a detector. By 1963 the project was finished, but the antenna remained.

Two American radio astronomers, Arno Penzias and Robert Wilson, commandeered the antenna, using it as a telescope to search for cosmic signals at a wavelength of 7.3 cm. In 1965, they discovered a faint noise signal, something like the hiss from an untuned TV, coming from all over the sky. The noise could have been caused by faults, electrical noise, or from the antenna itself. Two pigeons had nested inside the horn, and it was coated with a layer of droppings, which might have affected the results. The astronomers cleaned off the deposit and tried to get rid of the birds. But they kept on coming back, as pigeons tend to do. More drastic measures had to be taken, and this time the birds did not return. After painstaking elimination of all remaining possibilities, the noise refused to go away, and the two astronomers eventually accepted that it came from the cosmos.

The radiation they discovered has a temperature of 2.7 K, and turned out to be the missing piece of the jigsaw puzzle, the proof that the Universe had once existed in a very high-temperature state. Penzias and Wilson were awarded the 1978 Nobel Prize for the discovery of the CMB.

Soon after, a plethora of measurements were made in the 10–200 mm waveband from the ground and balloons. All

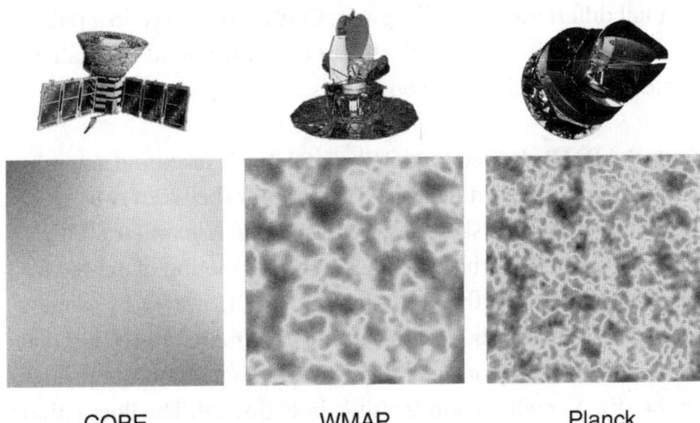

29. Three cosmic microwave background (CMB) satellites, showing the improvement in resolution of the CMB temperature fluctuations measured in the same area of sky.

confirmed a blackbody spectrum with this temperature. The spectrum actually peaks at a wavelength of 2 mm where the atmosphere is opaque. So, to prove definitively its primordial origin, the spectrum had to be measured from space.

In 1989 the *Cosmic Background Explorer* (COBE) satellite (Figure 29) was launched to do this. The satellite carried cooled bolometers in the *far infrared absolute spectrophotometer* (FIRAS), sensitive to wavelengths of 0.1–10 mm. The telescope mapped the whole sky, nearly twice in the ten months before the coolant ran out. The results (Figure 11), revealed the most perfect blackbody spectrum that has ever been measured, with a temperature of 2.725 K.

The second part of COBE's mission, measuring the smoothness of the CMB distribution over the sky, was done with a *differential microwave radiometer* (DMR), an instrument used to measure the intensity of radiation. With the DMR's two horn antennas (each with 7° resolution) pointing in different directions, small irregularities in the sky temperature distribution could be revealed as small differences in their signals. COBE was placed in a polar orbit, keeping the Sun and Earth out of sight, behind a radiation shield, as the satellite spun on its axis, scanning the sky.

First, the DMR showed there to be a relatively large and smoothly varying temperature difference (0.2 per cent) between two opposite poles in the sky, known as the *dipole asymmetry*. This pattern results from the 600 km per second motion of our local frame of reference with respect to the CMB. However, by probing deeper in COBE's data, things became even more interesting. The CMB showed a map of small temperature fluctuations, imprinted on the sky. A small section is visible in Figure 29. The fluctuations are tiny (1 part in 100,000), and are close to the smallest signals that COBE could detect.

The significance of the fluctuations

The tiny CMB fluctuations have a significance that is out of all proportion to their size. In its early high-temperature phase, the Universe consisted of photons and plasma particles, tightly coupled to each other. In this churning maelstrom, photons could not travel more than a short distance before being scattered by electrons in the plasma. As the Universe expanded, it cooled. After 380,000 years, it became cool enough for neutral atoms to form (the time of recombination). This was the time when the photons were decoupled from the particles, and were set free to travel through space unhindered. The CMB photons are therefore the most ancient ones we can see, dating back to a time when the Universe was very young.

Before the time of recombination, the Universe was a noisy place, filled with sound waves which compressed and rarefied the gas through which they propagated. Over time, and with the help of gravity, the compressed regions began to clump matter together into regions of higher density—the density fluctuations. Photons were scattered more frequently from these overdense regions, in fact so much so that when they eventually became decoupled from the matter, the photons took away 'memories' of their last interactions with matter: wrinkles on the surface of last scattering.

It was these ancient wrinkles that COBE first saw in 1992 as diffuse blobs on the sky. The blobs were the seeds—destined to grow into the clusters of galaxies and the stars that we now see. It is remarkable that we are able to look so far back in time and see our own beginnings. Stephen Hawking described the COBE result as 'the scientific result of the century, if not of all time'. For leading the COBE project, George Smoot and John Mather were awarded the 2006 Nobel Prize.

Interpreting millimetre and microwave measurements of the CMB is difficult. Confounding signals come from two natural sources: at low frequencies there is synchrotron radiation from local galaxies, and at high frequencies, thermal radiation from interstellar dust. These emissions can be identified and removed from the CMB signal by making observations at several different frequencies. In 2001, a new space telescope, the *Wilkinson Microwave Anisotropy Probe* (WMAP), was placed into L2 orbit. Its role was to increase the angular resolution and improve the quality of the data, by monitoring five wavebands. WMAP mapped temperature differences across the sky at wavelengths of 3–13 mm, with a pair of parabolic antennas that could resolve structures 13′ across. Advances in detector technology were incorporated. Instead of cooled bolometers, WMAP used the more stable *high electron mobility transistor* (HEMT) detectors. These had been developed for UHF devices (like cell phones). The first WMAP results were published in 2003, confirming the COBE result, but showing more details of the fluctuations (Figure 29).

Can still more be obtained from the primordial fireball, informing us what happened *before* recombination? Possibly. There is more to find out about the CMB—its *polarization*. According to the theory of *cosmic inflation*, the Universe expanded by an enormous factor a split second after the Big Bang. The immense distortions and wrenches in the fabric of space-time should have generated primordial gravitational waves which would have contorted the plasma, leaving, as evidence, a characteristic pattern imprinted on the scattered photons. The theory of inflation is not proven, but if it is correct the CMB photons we now observe ought to show a special type of polarization, called a *B-mode* (resembling a twisted magnetic field).

To refine still further the CMB measurements, and to look for any polarization, the *Planck Space Telescope* was placed in L2 orbit in 2009. It monitored nine frequency bands, using a range of cooled instruments—HEMTs for 30–70 GHz, and bolometers for 100–837 GHz—and had higher resolution and more sensitive detectors to measure the intensity and polarization. The coolant for the bolometers was exhausted in 2012 and that for the HEMTs less than a year later.

The CMB results have profound implications for cosmology. The *standard cosmological model* is known as ΛCDM (or *lambda cold dark matter*). It contains three main ingredients: baryonic matter, cold dark matter (CDM, the unknown form of gravitating matter), and the mysterious *dark energy*. The dark energy (Λ) ingredient refers to a mathematical term in Einstein's cosmology equations, and represents a ubiquitous and dominant form of energy. Normal gravitational attraction tries to slow down the expansion of the Universe by pulling massive objects like galaxies together. But dark energy is an anti-gravity substance, responsible for the observed acceleration of the expansion of the Universe. As American physicist Alan Guth described it: if you throw a ball up in the air, instead of it returning to Earth, it just keeps accelerating away from you.

The ΛCDM model fits closely with the Planck telescope data, yielding the following cosmological parameters: the time since the Big Bang is 13.8 ± 0.037 billion years; the Universe consists of 4.9 ± 0.05 per cent baryonic matter, 26.8 ± 0.4 per cent CDM, and 68.3 ± 1 per cent dark energy.

The multi-coloured Universe

When telescopes operating at different wavelengths are used together, they become extremely powerful. The benefits were recognized in NASA's great observatories programme at γ-ray, X-ray, UV, visible, and IR wavelengths. The power of the multi-frequency approach was already apparent in the 1960s when radio and optical telescopes joined forces, leading to the discovery of quasars.

The famous Crab nebula (M1) is the remnant of a supernova explosion that was observed worldwide, and recorded by Chinese astronomers on 4 July 1054. The Crab has been observed at all wavelengths: from radio to γ-rays. Its origins lay in the core of a massive star, depleted of fuel, which could no longer support itself against gravitational collapse. The star exploded, leaving the core as a neutron star 10 km across. The star's outer shell was violently ejected at a speed of 1,000 km per second, ploughing into the surrounding interstellar medium, and forming the gaseous nebula—a tangled mass of glowing filaments, now about ten light years across.

The nebula emits polarized light, with a luminosity of 75,000 Suns, and the stellar remnant, the pulsar, spins thirty times a second. As its powerful magnetic field sweeps round, electrons are accelerated close to the speed of light and broadband pulses of EM radiation beamed out. A fierce gale of energetic particles, a *pulsar wind*, blows out, filling the nebula and generating synchrotron radiation as particles are trapped, spiralling around magnetic fields.

At visible and UV wavelengths (Figure 30), HST images of the core show complex gas motions and glowing tendrils. In the IR,

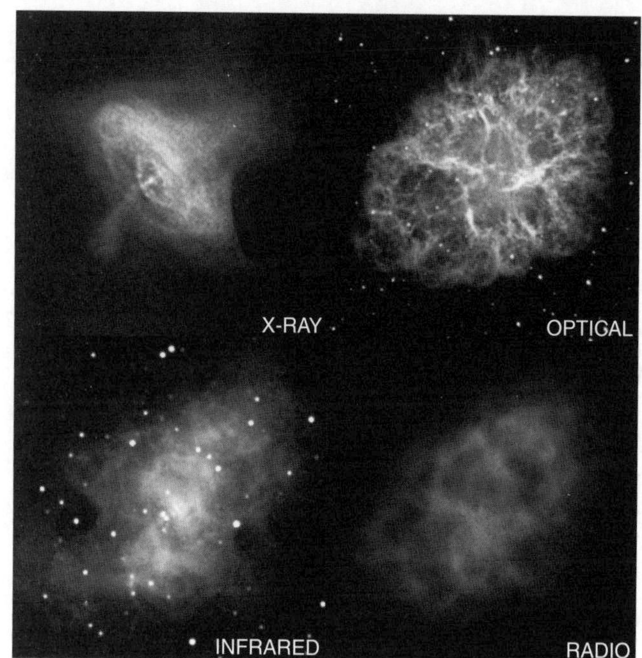

30. The Crab nebula at different wavelengths.

Spitzer shows cooler dust-laden filaments that were ejected in the explosion, embedded in a cloud glowing with synchrotron radiation, which is visible at radio wavelengths. But the most dramatic images are at high energies: X-rays and γ-rays. Chandra reveals the Crab to be highly dynamic, and Fermi's observations show it to be one of the brightest point-like γ-rays sources in the sky, with highly variable bursts of 100 MeV photons.

The Crab pulsar is gradually spinning down. The pulse rate has been monitored for decades, clearly revealing its slowing heartbeat. The pulsar is losing rotational energy, first converting it to particle energy, and then to EM radiation. The numbers match: the rate at which rotational energy is lost balances the luminosity of the nebula. The rotating neutron star is, effectively, a power generator—a much more extreme version of the rotating armature of an electric generator.

The last seven decades have seen telescopes launched into space, vastly enhancing the crispness of the images they produce and expanding the range of observable wavelengths. Space telescopes have transformed our knowledge. The information we have gleaned from them has enabled us to make new discoveries and form much more complete astrophysical models. However, each question that telescopes have helped answer has led to new questions. The mystery deepens at each step of the way. To address the new questions, we are now building a new generation of telescopes.

Chapter 9
The next telescopes

The key questions in astronomy today are: *what is dark matter*; *what is dark energy*; *how did the first galaxies form*; and *are there habitable, Earth-like exoplanets*? The last question includes the search for *biomarkers*—signs of the molecules of prebiotic life. The next decade will see powerful new telescopes come into operation to address these questions. These are today's questions. Tomorrow's questions may well be different, and, if there is anything we can learn from the history of telescopes, they will be. We must therefore allow for serendipitous discoveries; as we have seen, astronomy is littered with these.

Size matters, and so does resolving power

In 2009, the telescope celebrated its 400th anniversary. For four centuries, the collecting area has increased exponentially. The largest optical telescope planned to start operating in the next decade will collect a hundred million times more light than the eye. Today's large telescopes have already detected some of the faint light emitted by highly redshifted galaxies from the dawn of the cosmos, infant galaxies that existed when the Universe was less than a billion years old.

But it is a struggle. The enemy of weak signals is *noise*, the random contributions from various sources that can easily overwhelm signals arriving at a detector. The act of counting photons or photoelectrons introduces a statistical uncertainty called *shot noise*, which is related to the quantum graininess of light and matter, and becomes worse the fewer the photons there are to count. Noise can also come from the sky near the target and from thermal radiation produced in the detector itself.

To maximize the probability of detecting a weak signal, a telescope needs to collect as many photons from the target as possible, and concentrate them into the smallest possible spot. The flux of photons collected by a telescope of aperture D is proportional to D^2, and, if the telescope is diffraction-limited, the diameter of the image of a point source is proportional to λ/D. The area of the image spot therefore varies as $1/D^2$, and the flux there is proportional to D^4.

This steep dependence underlines the importance of a telescope having a large aperture *and* a high resolution, a successful combination that is well borne out by the Hubble Space Telescope (HST), with its diffraction-limited resolution and 2.4-metre mirror. With the advent of adaptive optics systems, large ground-based telescopes can now approach their diffraction limit, thereby greatly increasing their performance.

This logic underpins the designs of three extremely large IR/optical telescopes (ELTs), equipped with adaptive optics systems, due to start operating in the next decade. The largest of these, the European Southern Observatory's (ESO) *European Extremely Large Telescope* (E-ELT, Figure 31), has a 39-metre (f/1) segmented mirror, and is to be built on Cerro Armazones in Chile (altitude 3,060 m). It is expected to achieve its diffraction limit (0.01″, at a wavelength of 2 μm). There will be two Nasmyth foci (f/17), on platforms the size of tennis courts placed each side of the telescope, and the whole moveable structure will weigh 2,700 tonnes. The E-ELT has a five-mirror optical design,

31. Artist's impression of the European Extremely Large Telescope (E-ELT). Note the relative size of cars.

including a 6-metre secondary mirror and two flat mirrors in the main optical train; adaptive optics corrections will be made one thousand times a second. The field of view is 10′, one-third of the width of the full Moon.

The E-ELT is in fact a scaled-down version of the concept for a 100-metre telescope—the OWL, or *Overwhelmingly Large Telescope*. The OWL will not be built, for budgetary reasons, at least for now.

Two other ELTs planned are the *Thirty-Metre Telescope* (TMT), which will use a segmented mirror, planned to be built on Mauna Kea, Hawaii, and the *Giant Magellan Telescope* (GMT), which will use seven 8.4-metre honeycomb mirrors. The GMT is being built at Las Campanas Observatory in Chile, and will have an effective collecting area equivalent to one 22-metre telescope. Each of the three ELTs will have a comprehensive suite of instruments for imaging, photometry, and spectroscopy from the optical to the IR.

Making movies of the sky

The 8.4-metre optical and near-IR *Large Synoptic Survey Telescope* (LSST) was approved in 2014 and will be built on Cerro Pachon peak (altitude 2,175 m) in Chile. Observations are due to start in 2022. There are several novel features about this robotic telescope, not least its three-mirror design to obtain a very wide field of view (3.5°, equivalent to fifty full Moons). The three mirrors are: an 8.4-metre primary, a 3.4-metre secondary, and a 5.2-metre tertiary, formed out of the same glass blank as the primary. The small focal ratio (f/1.25) is achieved by a compact design and multiply folded light path. The camera, a 3.2-gigapixel mosaic of 200 CCD arrays, will be the largest digital camera ever built—the size of a car. It will give an exceptionally uniform field of view across the large 64 cm image plane, corrected by refractive elements. The telescope will use optical and IR filters in six wavebands, to determine redshifts photometrically. Thirty terabytes of data will be collected each night, feeding into a database that will total 60 petabytes.

The LSST will survey the same half of the southern sky (10,000 square degrees) every three days. This high cadence will enable it to make 'movies' of the sky—with this frame rate. The databases will contain about 20 billion galaxies, and a similar number of stars. The telescope is expected to discover many transient objects, for example: supernovae, potentially hazardous asteroids and near-Earth objects, GRBs, and variable stars. Objects that change position or brightness will trigger alerts for other observatories to follow up within a minute of their discovery, and this will include new type 1a supernovae.

The LSST will be well placed to measure the weak gravitational lensing of dark matter. Because the telescope will see fainter and more distant galaxies than, for example, the SDSS, its deep survey catalogue will contain higher redshift galaxies. From their weak gravitational lensing properties, the LSST will provide data to create a map of dark matter structures, particularly showing the time evolution. This 3D map can, in turn, also be used to infer the properties of dark energy.

In space, the *Wide-Field Infrared Survey Telescope* (WFIRST), planned for the mid-2020s, will, like the HST, use a 2.4-metre mirror but will cover one hundred times the area at wavelengths of 0.6–2 μm. The observatory will survey many galaxies and clusters of galaxies, and will also investigate the distribution of dark matter via gravitational lensing. WFIRST will also use gravitational microlensing to look for planetary signatures. If the lensing star has a planetary system, there will be measurable deviations from simple microlensing behaviour, which will yield information on the planets. The *Euclid* (1.2 m) optical/NIR space telescope, due for launch in 2020, will measure the expansion of the Universe out to a high redshift ($z = 2$) with great precision. To do this Euclid will survey a billion galaxies, producing a vast amount of data, which will be used to make a map of the 3D distribution of galaxies, dark matter, and dark energy. This will map the evolution of cosmic structure over the last 10 billion years.

The Square Kilometre Array

The *Square Kilometre Array* (SKA) is a radio telescope with a collecting area of a million square metres. It will soon be the world's largest scientific instrument, and one that will transform our view of the Universe. The SKA will expand the sensitivity, resolution, and spectral coverage of existing radio telescopes by more than an order of magnitude. The frequency coverage is wide: from 50 MHz to 20 GHz, spanning the low frequencies of LOFAR up to the high frequencies of ALMA. When completed, in the mid-2020s, the telescope will bridge two continents: from the Karoo region of South Africa, to the Murchison region of Western Australia. The locations have been selected on the basis of their sparse populations and lack of radio frequency interference—radio quietness is vital to the success of this project. There are three elements to the telescope: SKA-mid, SKA-survey, and SKA-low, reflecting the different techniques needed to cover such a broad spectrum. The undertaking is so large that its various parts will come online in a stage-managed and carefully choreographed manner.

SKA-mid, in South Africa, will consist of a large-aperture synthesis array of 250 15-metre paraboloidal radio dishes operating between 350 MHz and 15 GHz. The antennas will be arranged in three groups. The most compact of these is a 5 km dense 'core' or cluster of antennas, containing half of the total collecting area. The core will yield high sensitivity, for surveys. Outside the core are intermediate antenna groupings, extending to distances of 180 km. An outer array, along winding spiral arms, will have long baseline stations with antennas located out to 3,000 km. These widely separated antennas will yield information on milliarcsecond angular scales. The finished array will incorporate a smaller pathfinder telescope array (MeerKAT), already being constructed, and producing initial results in 2016.

The *SKA-survey* telescope, in Australia, will use the same type of dishes as SKA-mid, but will be equipped with *phased-array feeds* (PAFs). In a PAF, the focal plane of a dish is filled with an array of antenna elements, individually connected to a central site. The elements will be controlled via software to form a phased array, creating a pattern of radio beams and viewing different parts of the sky all from one dish. The use of PAFs will therefore greatly speed up sky surveys. In the *Australian Square Kilometre Array Pathfinder* (ASKAP) radio telescope, the PAF concept is currently being tested as part of an array of thirty-six 12-metre dishes at frequencies between 0.7 GHz and 1.8 GHz. ASKAP will have thirty-six beams per dish and will cover a 30 square degree area of sky.

SKA-low (also in Australia) will be an aperture array, with 250,000 antenna elements for low frequencies (50–350 MHz).

There are no moving parts; each antenna element is able to see the whole sky. The antennas are connected using fibre optic cables (with a total length able to wind twice around the Earth). The antennas will be phased to form steerable beams, using software control. Like SKA-mid, most of the antenna elements are densely packed in the central 2 km of the array, and there are also others 50 km away on spiral arms, providing high-resolution information. The concept is that of a software telescope, currently being pioneered by LOFAR.

The scientific goals of the SKA revolve around the questions posed at the start of this chapter. However, new fields will be opened up. For example, the SKA will be able to probe the cosmic dawn of the Universe, using 21 cm neutral hydrogen radiation, strongly redshifted to a wavelength of a few metres.

The *Cradle of Life* project will utilize the SKA's radio spectroscopy capability to search for the radio lines of prebiotic molecules, the building blocks of life, such as amino acids. The large collecting area of the SKA could detect faint extra-terrestrial signals, if they exist, which would expand on the *Search for Extra-Terrestrial Intelligence* (SETI) project. The SKA's antennas will be sensitive enough to detect an airport radar transmitter on a planet fifty light years away. Within this distance there are over one hundred stars similar to the Sun, many of which may possess Earth-sized planets.

The large collecting area of the SKA also makes it well suited to discovering faint transient sources—such as pulsars in galaxies. The number of known pulsars is around 2,000, a figure that is likely to increase ten times with the SKA. It is hoped that discoveries of new binary pulsars, orbiting in the strong gravitational fields of black holes, will reveal fundamental information on the fabric of space and time. Precise timing measurements of pulses from a large ensemble of pulsars should reveal the presence of gravitational waves produced by merging black holes, and intersect with the direct detection of these gravitational waves with instruments like LIGO.

The SKA may also be able to shed light on a poorly understood subject: the origin of *cosmic magnetism*. It will be able to do this by observing the evolution (with redshift) of the polarization of synchrotron radio emission from galaxies.

The James Webb Space Telescope

Some of the most important of Hubble's results are the images of the Hubble deep fields. However, the expansion of the Universe limits what can be seen of the first galaxies when viewed in visible light. The UV light from very early galaxies has been redshifted right through the visible, and into the mid- and far-IR regions. In effect, an IR telescope is needed to look at the edge of the Universe and do UV astronomy. This is the role that the next large space telescope will fill.

Twenty-five years after the launch of the HST, we are preparing for an Ariane 5 rocket to launch its successor, the *James Webb Space Telescope* (JWST) in 2018. The JWST is a wholly IR telescope with a huge 6.5-metre mirror, collecting seven times the light of the HST. Webb will also carry more sensitive cameras and spectrographs (including an IFS) for mid- and near-IR wavelengths (0.6–28 μm), making it one hundred times more powerful than the HST.

The lightweight gold-coated JWST mirror is made from a mosaic of eighteen light, strong beryllium hexagonal segments that are folded up on launch, like origami. The folding procedure is necessary to squeeze such a large mirror into the nosecone of the rocket. When the telescope arrives at its L2 orbital point the segments will unfurl, like a butterfly emerging from a chrysalis, to produce the paraboloidal mirror, supported on a backplane structure (Figure 32). The mirror has to be very cold (53 K)

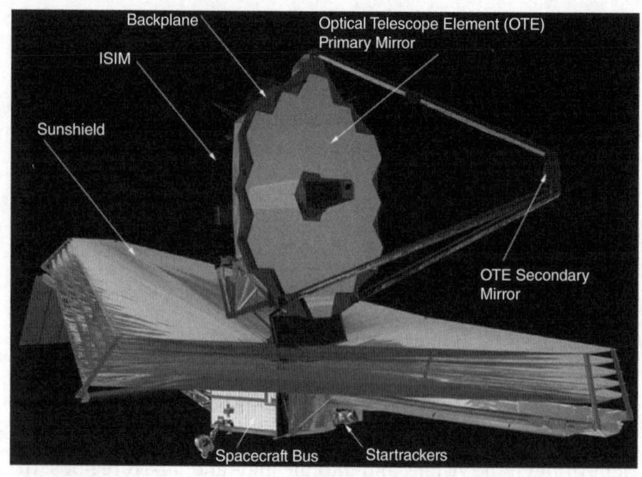

32. The 6.5-metre James Webb Space Telescope mirror sitting on its sunshield.

so that its own heat radiation does not overwhelm the faint cosmic light. The telescope's detectors, located in the integrated science instrument module (ISIM), also need to be cooled and the telescope will have a sunshield, the size of a tennis-court, blocking radiation from the Sun, Earth, and Moon.

At IR wavelengths, Webb will also be able to probe interstellar dusty star-forming clouds for protoplanetary discs, observe emission from organic molecules, and see inside star-forming regions, such as the *Pillars of Creation*—but with one hundred times the resolution of the HST. The telescope will also search for the direct light from exoplanets and image them with a *coronagraph*, a disc placed in the focal plane to block the light from the star. The telescope will be used to search for biomarkers—the IR spectral signatures that include oxygen, carbon dioxide, and water vapour molecules in the atmospheres of exoplanets.

Flashes in the sky

Although the atmosphere blocks wavelengths shorter than UV, the most energetic γ-ray photons (above 10^{11} eV) interact with the molecules above 10 km in the atmosphere to generate visible light flashes, which can be observed with ground telescopes. The pair-production process, involving the close-collision of a photon with a nucleus, results in a cascade of secondary particles, an *air shower* (see Figure 33), producing a brief, narrow, and forward-pointing beam of photons, illuminating a 250-metre diameter patch of ground. This light is *Cherenkov radiation*, the same as the bluish light produced by high-energy particles in a water tank surrounding a fission power plant. It is generated by the fast particles exceeding the local light propagation speed, and can be thought of as being equivalent to the 'boom' of a supersonic aircraft.

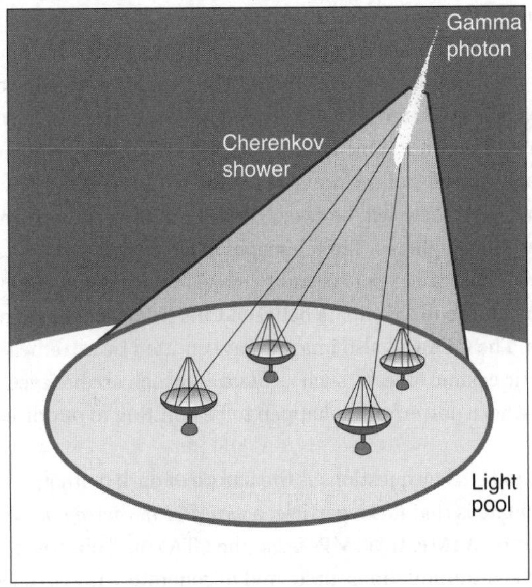

33. An Imaging Atmospheric Cherenkov Telescope.

The flash is observed with arrays of optical telescopes—*Imaging Atmospheric Cherenkov Telescopes* (IACTs), first appearing in the 1990s. One observatory, the *High-Energy Stereoscopic System* (HESS, 2012) in Namibia, has six optical telescopes (five with 12-metre mirrors and one with a 28-metre mirror) spaced 120 metres apart on the ground. The telescopes are used to view an air shower from different perspectives, and so reconstruct its 3D geometry. The images can be back-projected, pointing to the γ-ray source in the sky. Objects emitting such high-energy γ-ray photons include supernovae and supernova remnants, the nebulae surrounding pulsars, γ-binary systems, massive stellar clusters, and AGNs. More than ninety sources producing 10^{12} eV γ-rays have been discovered with HESS so far.

The next large IACT, ESA's *Cherenkov Telescope Array* (CTA, 2020), will increase the aperture and resolution tenfold using a hundred telescopes, with apertures ranging from 4 to 23 metres. Measuring Cherenkov flashes is challenging: the flashes last for less than a few hundred nanoseconds, and images moves rapidly across the focal plane of a telescope. The flashes are much weaker than background starlight and airglow, and must be recorded to within nanoseconds. One issue is dealing with the large quantity of data produced—of the order of petabytes (10^{15} bytes). To cover the whole sky, there will be two CTA sites: in the northern and southern hemispheres. Ten telescopes at a time will record more than a billion frames per second to make movies of the radiation showers and so discriminate between Cherenkov and background signals. The CTA will also image γ-rays emitted by extremely energetic cosmic objects, such as *blazars*, which are believed to be AGNs whose powerful jets happen to be pointing in our direction.

One of today's key questions is the nature of dark matter. A candidate is that it is a particle, a *weakly interacting massive particle*, or WIMP. If WIMPs exist, the CTA could observe their mutual self-annihilations, predicted to generate γ-rays of specific energies. So far, high-energy γ-ray photons have been seen only to come from discrete point-like sources. But, if dark matter exists clumped in substructures in the halos of clusters of galaxies, the CTA could reveal their existence by detecting their γ-rays.

Big data

Figure 14 is very telling: it shows that the light-collecting area of telescopes has doubled every thirty years, however *nothing* has driven the quantity of data produced by telescopes as hard and as fast as digital computer technology. This is exemplified by the steep growth, from the 1980s, of the number of pixels in a CCD camera, increasing at least as fast as Moore's law, doubling about every two years. Before the digital age, astronomers were limited by the capacity of telescopes and their instruments to collect data. Today we are in danger of being overwhelmed by a tsunami of it.

Telescopes coming online in the next decade will feed the voracious appetites of enormous multidimensional databases. To see how this has happened, let's start with small data and see where that leads us. Building up from the simplest data structures, a one-dimensional list of numbers could be, for example, a spectrum such as the brightness of a galaxy as a function of wavelength, or a time-series describing the brightness of a star with a transiting exoplanet. In two dimensions the data might be an image (for example, sky brightness expressed in (x, y) coordinates, in a CCD camera). A CCD image with 2,000 × 4,000 pixels contains 16 MB of data. Adding another dimension to this, for example wavelength, produces a three-dimensional *data cube* (x, y, λ).

A data cube can be produced by an IFS in which the light arriving at each pixel might be dispersed into a large number, say, 50,000 spectral points or more. The storage needed for such a cube would be at least a terabyte (TB, 10^{12} bytes). The third dimension could instead be time, in which case the cube would represent a movie, showing transient events in one patch of sky. A sequence of IFS data cubes might also be taken at various times, resulting in a four-dimensional object—a time-series of cubes (a three-dimensional movie). The data explosion will be even greater with the recent development of detectors such as the TES devices, which, like CCD X-ray detectors, give counts that are proportional to the incident photon energy, yielding simultaneous spectral, time, and positional information.

From this, it is easy to see how small data can quickly become big data. Storage, however, is not the main issue; it is the size and complexity of the data. For the LSST or SKA telescopes, the data traffic will compete with that of the worldwide web as it is today. In 2016 this was estimated to be of the order of 1 zettabyte (ZB, 10^{21} bytes)—and growing. It would take 36 million years to watch one ZB of high-definition streaming video. This is the size of the data 'haystack' the SKA will produce in one year. This quantity exceeds the amount that can be examined by all the world's astronomers.

In this haystack there are of course needles to be found, the scientific discoveries. The question is: how best to find them? There are three possible answers. One is to make use of *crowdsourcing*—the eyes and brains of millions of citizen scientist volunteers have already classified SDSS objects in various Zooniverse projects. Many of the galaxies viewed and classified online by members of the public had never previously been seen by human eyes. Millions of successful classifications have been made this way, and stringent crosschecks for consistency have been applied. However, the eyes and brains of the volunteers will only be able to classify a small fraction of the data that is about to be unleashed.

Another way forward is to develop advanced web browsing software, for example Microsoft's *WorldWide Telescope* (WWT). The WWT is a data visualization system that combines archived images, spectral data, and surveys from X-ray to radio wavelengths, and from very many telescopes. The data is open source (freely available in the public domain) and stored on servers in 'the cloud'. Anybody can chart a voyage through the virtual Universe and zoom in on features of interest in any part of the sky, combining them with live views from telescopes and look at all the relevant scientific publications. All these data are put into a common format so that it is possible to obtain a holistic view, exploring polychromatic, multidimensional data in new and original ways, and cross-linking with topics of shared interest. The WWT is designed to expand flexibly, as data from new telescopes are gathered. The platform is also designed for a range of users, from early educational explorers to professional research astronomers. It is likely that those who can most dexterously perform the highly complex manipulations in a multidimensional space will discover the 'needles in the haystack'.

There is a third possible way to find the needles: apply autonomous *pattern-recognition* (machine learning) computer applications to search out and classify features in the data. Allied with this approach is the use of crowdsourcing data to *train* search algorithms how to classify objects. Such methods require large amounts of computer processing power, and this approach is considered feasible for the next decade using computers with multiple parallel processors. Any objects of unusual interest found by *data automatons* can of course be displayed to human eyes for validation.

Epilogue

Our journey through four centuries of telescopes has taken us from the simple spyglass, little more than a toy consisting of two lenses in a tube, to the Square Kilometre Array, shortly to be the world's largest scientific instrument. In the hands of a genius, Galileo's 'toy' showed breathtaking views of the cosmos, and precipitated the collapse of a 2,000-year-old cosmology that had placed the Earth at the centre of the Universe.

After Galileo, our knowledge of the heavens accelerated. Telescopes got bigger and were combined with spectrographs and photographic instruments. The chemical elements in the stars were shown to be the same as those on Earth. Only a hundred years ago, when Einstein published his general theory of relativity, we knew what chemical elements were present in the Sun (but not how they came to be made in stars), we knew the distances to a handful of the nearest stars (but not the distances of the spiral nebulae, or indeed if they were part of the Milky Way). We had no idea we live in an expanding Universe, let alone one that started with a Big Bang 13.8 billion years ago. Telescopes have been the keys that unlocked this knowledge.

Today, even larger optical and infrared telescopes are now seeing very young galaxies, less than a billion years after the Big Bang, which appear very different from galaxies today. We have mapped the oldest light, the CMB radiation, and found minute temperature fluctuations in it that were destined to form the clusters of galaxies, galaxies, and stars that now fill the observable Universe. Information from across the whole of the electromagnetic spectrum, from γ-ray to radio wavelengths, has shown the existence of exotic gravitationally collapsed objects, black holes and neutron stars, and the energetic astrophysical processes they engender. We have discovered that exoplanets are common. We have caught sight of protoplanetary discs forming new solar systems in stellar nurseries, embedded inside giant molecular clouds.

The next decade will see powerful new telescopes coming on line. With these, we will be uniquely placed to investigate deep mysteries: the nature of dark matter, dark energy, what happened during the dark age of the Universe, and how this era came to an end when the first stars shone out brilliantly, heralding the cosmic dawn. New telescopes will be trained on exoplanets and their atmospheres, searching for biomarkers—the molecules that could indicate the presence of extra-terrestrial life.

It is appropriate to end this *Very Short Introduction*, at least for now, with one final message. If there is one thing that we can learn from history, it is that we should expect the unexpected. As the eminent physicist Neils Bohr put it: 'Prediction is very difficult, especially if it's about the future.'

Further reading

K. Blundell, *Black Holes: A Very Short Introduction* (Oxford University Press, 2015).

R. Florence, *The Perfect Machine—Building the Palomar Telescope* (Harper Perennial, 1995).

T. Hey, S. Tansley, and K. Tolle, Eds, *The Fourth Paradigm: Data-Intensive Scientific Discovery* (Microsoft Research, 2009).

H. C. King, *The History of the Telescope* (Dover, 2003).

M. Longair, *The Cosmic Century* (Cambridge, 2006).

I. S. McClean, *Electronic Imaging In Astronomy: Detectors and Instrumentation*, 2nd ed. (Springer, 2010).

W. P. McCray, *Giant Telescopes* (Harvard University Press, 2004).

F. G. Smith, *Eyes on the Sky* (Oxford University Press, 2016).

F. G. Smith and J. H. Thomson, *Optics* (Wiley, 1973).

I. A. Walmsley, *Light: A Very Short Introduction* (Oxford University Press, 2015).

Publisher's acknowledgements

We are grateful for permission to include the following copyright material in this book.

The quotation on p. 19, from Dante Alighieri's Inferno from the *Divine Comedy*, Canto XXXIV, line 139: 'E quindi uscimmo a riveder le stelle', translated by Richard Dunn and used with permission.

The publisher and author have made every effort to trace and contact all copyright holders before publication. If notified, the publisher will be pleased to rectify any errors or omissions at the earliest opportunity.

Index

21-cm hydrogen line 91–2, 128
51 Pegasi b 75

A

aberration 2
  coma 21
  chromatic 21, 28
  spherical 21, 24, 67, 78, 105
adaptive optics (AO) 46, 70ff., 86, 123
achromatic doublet 4, 28
active galactic nucleus (AGN) 84, 98, 101, 104
airglow 44, 51, 132
altitude-azimuth (altazimuth) mount 27–8, 65–6
Andromeda galaxy (M31) 8, 25, 30, 32, 36, 48–9
Angel, Roger 67
Anglo-Australian Telescope (AAT) 5, 73
angular resolution 16–18, 46, 66–7, 69, 72, 78–9, 84, 86–7, 94, 103, 105, 117
aperture synthesis 84ff., 89, 94, 127
Arecibo Observatory 78, 90
astronomical unit (AU) 30
astrophotograph 50
Atacama Large Millimetre Array (ALMA) 94–5, 126
Australian Square Kilometre Pathfinder (ASKAP) 127

B

Baade, Walter 36–7, 89
Babcock, Horace 70
baryon 9, 101, 109, 118–19
Bell, Jocelyn 88–9
Beta Pictoris b 69
binary pulsar 90, 128
Big Bang 5, 35, 61, 107, 109, 113, 118, 136
big data 3, 5, 133ff.
biomarker 5, 122, 130, 136
black hole 5, 91, 96, 99–101, 107, 128–9, 136
  supermassive (SMBH) 69, 84, 87–8, 103, 106
blackbody radiation 42ff., 83, 87, 96, 115
blazars 132
bolometer 55–7, 94, 115, 117–18
Bond, William 50
Boyle, Willard 52
Bunsen, Robert 59

C

catadioptric 24
Cassegrain 22–4, 31, 37, 66, 69, 105
cepheid variable 31–2, 35–6, 106
celestial sphere 25–7
Chandra X-ray Observatory (CXO) 58, 102–3, 120
charge coupled device (CCD) 48, 52ff., 58, 73, 76, 103, 109, 125, 133–4
Cherenkov radiation (also Cherenkov Telescope Array; CTA; IACT) 131ff.
Chester Moore Hall 28
chopping and nodding (a telescope) 55
clusters, globular 31
  of galaxies 96, 101, 108, 117, 126, 133, 136
  of stars 25, 31–2, 50–1, 56, 132
coded aperture mask telescope 99–100
Compton, Arthur Holly 57
  Gamma Ray Observatory 98
  scattering 57–8
  scattering, inverse Compton 97
Copernicus satellite 104
corona, solar 80–3, 112
coronagraph 130
Cos-B satellite 97
Cosmic Background Explorer (COBE) 43, 115ff.
cosmic dawn 92, 128, 136
cosmic inflation 117–18
cosmic magnetism 87, 129
cosmic microwave background (CMB) radiation 43, 92, 113ff.
Crab nebula (M1) (and pulsar) 25, 98, 101, 119ff.
Cradle of Life Project 128
Curtis, Heber 32

D

Daguerre, Louis-Jacques-Mandé 50
dark age (of Universe) 92, 136
dark energy 70, 118–19, 122, 126, 136
dark matter 36, 101, 111, 118, 122, 125–6, 132–3, 136
data cube 73, 133–4
De la Rue, Warren 50
Deep (and Ultra Deep) HST fields 106–7, 129
deformable mirror 70–1
Digges, Leonard and Thomas 19
diffraction 16ff., 46–7, 82, 105, 123
  grating 60ff.
dish antenna 77–9, 81, 83–6, 88, 93–4, 127
dispersion, of light 20–1, 54, 59, 62, 73, 133
Dobson, John 27
Dollond, John 29
Doppler, Christian 34
  shift 34, 75, 90
  spectroscopy 74, 91, 106
Draper, John William 50

E

Earth rotation synthesis 85
Einstein, Albert 11, 57, 90, 98, 103, 107, 118, 136
  X-ray Observatory 101
electromagnetic field 40ff.
  spectrum 1, 41ff., 96, 136
electron volt (eV) 11
equatorial mount 26–8, 37
Euclid satellite 126
European Extremely Large Telescope (E-ELT) 64, 123–4
European Southern Observatory (ESO) 67, 123
European Space Agency (ESA) 97–8, 105, 110, 113, 132

Event Horizon Telescope 88
exoplanet 5, 69, 74–6, 109–12, 122, 130, 133, 136
expansion of Universe 33ff., 38, 66, 70, 92, 106, 113, 118, 126, 129
extra-terrestrial life, search for 5, 49, 128
eyepiece 15–16, 22–4

F

field of view 24, 53, 72, 74, 76, 109, 124–5
Fermat, Pierre de 12
Fermi Gamma Ray Telescope 100, 120
focal length 13–16, 21, 28, 37, 81, 103
focal ratio 21, 24, 125
Foucault, Léon 29
Fourier, Joseph 82
  components 82ff.
four-acre array radio telescope 88
Fraunhofer, Joseph 28–9, 59
  lines 59

G

Gaia satellite 110
Galaxy Evolution Explorer (GALEX) satellite 104
Galileo, Galilei 1, 3, 5, 19–20, 29, 32, 64, 135–6
gallium-germanium (Ga-Ge) detector 56
gamma ray 2, 97ff., 131
  burst (GRB) 74, 97ff., 125
Geiger-Müller tube 57
Gemini telescopes 67–8
general theory of relativity 90, 103, 107, 136
geometrical optics 12, 107
Giant Magellan Telescope (GMT) 64, 124
Gran Telescopio Canarias (GTC) 66, 68
gravitational lensing 107ff.
  microlensing 111, 126
  weak 109, 125–6
gravitational waves 90–1, 118, 128–9
grazing incidence X-ray telescope 101–3
great debate, the 32
Gregorian 22

H

Hadley, John 27
Hale, George Ellery 22, 29, 36
  60-inch telescope 31
  100-inch telescope *see* Hooker
  200-inch Hale telescope 37ff., 63–8, 79, 84
Harriot, Thomas 19
Hazard, Cyril 83
Henry, Paul and Prosper 51
Herschel, John 50
Herschel, William 24–5, 29, 54, 64
  Space Observatory 113
Hertz, Heinrich 8, 11, 41
Hewish, Anthony 88–9
Hey, James 78
high electron mobility transistor (HEMT) 117
High Energy Stereoscopic System (HESS) 132
Hipparcos satellite 110
Hobby-Eberly Telescope (HET) 66–8
Hooker (100-inch) Telescope 32ff., 36, 64, 80
hot Jupiter, a 75, 112–13
Hubble, Edwin 32ff.
  constant (and law) 35ff., 73, 113

Hubble Space Telescope (HST) 44, 104ff., 119, 123, 126, 129–30
Ultra Deep fields 106–7, 129
frontier fields 109
Huggins, William 54
Hulse, Russell and Joe Taylor 90
Humason, Milton 35
Huygens, Christiaan 12, 22, 60–1

I

Imaging Atmospheric Cherenkov Telescope (IACT) 132
indium-antimonide (In-Sb) detector 56
Infrared Astronomical satellite (IRAS) 112
integral field spectrograph (IFS) 62, 73, 129, 133
interference 9–10
fringe visibility 80
fringes 9–10, 80ff.
interferometer 66–7, 69, 79ff.
International Gamma Ray Astrophysics Laboratory (INTEGRAL) 98–9
International Ultraviolet Explorer (IUE) 104
ionization chamber 57
ionosphere 39, 45, 77

J

James Clark Maxwell Telescope (JCMT) 93
James Webb Space Telescope (JWST) 129ff.
Jansky, Karl 3, 77

K

Karl G. Jansky Very Large Array (JVLA) 86ff., 95
Keck telescopes 64, 65ff., 68, 72
Kepler satellite 109ff.
Keplerian telescope 15
Kirchhoff, Gustav 59

L

Lagrange, Joseph-Louis 110–11
lambda CDM (ΛCDM), the standard cosmological model 118–19
Langley, James Pierpoint 55
Large Binocular Telescope (LBT) 64, 67–8
Large Synoptic Survey Telescope (LSST) 53, 64, 125ff., 134
Las Cumbres Observatory Global Telescope Network 76
Laser Interferometer Gravitational-Wave Observatory (LIGO) 90, 129
Leavitt, Henrietta Swan 31
lenslet 62
Lippershey, Hans 19
Liverpool Telescope 74
Lockyer, Norman 60
long-slit spectrograph 62
Lovell Telescope 78–9
Low Frequency Array (LOFAR) 5, 92–3, 126
Lowell, Percival 49

M

Mather, John 117
Maxwell, James Clark 40ff.
Magellanic Clouds 31–2, 36, 49, 104
magnetar 103
magnification 15–16
MeerKAT radio telescope 127
mercury-cadmium-tellurium (Hg-Cd-Te) detector 56
Messier, Charles 25
Michelson, Albert and Francis Pease 80

micrometer 49
millimetre wavelength astronomy 41, 93ff., 96, 117
mirror, meniscus 65, 67–8, 124
honeycomb 65, 67–8
segmented 3, 65–6, 123–4
spherical 22–3, 66–8, 78
Mullard Radio Astronomy Observatory (MRAO) 85, 89
Multi-Element Remote-Linked Interferometer (MERLIN) 85–6
Multiple Mirror Telescope (MMT) 64–5

N

NASA's four great observatories 98, 102, 104, 112
NASA's four Orbiting Astronomical Observatories (OAO) satellites 104
Nasmyth, James Hall 23
focus 23, 26, 66–7, 69, 123
Nelson, Jerry 65
Neugebauer, Jerry and Robert Leighton 55
neutral hydrogen 21-cm line 91–2, 128
neutron star 5, 89–90, 98, 100, 119, 121, 136
New Technology Telescope (NTT) 67
Newton, Isaac 3, 22, 59
Newtonian telescope 22, 23–4, 25, 27, 64
nodding 55
noise 54, 72–7, 112, 114, 122–3
Nuclear Spectroscopic Telescope Array (NuStar) 103

O

Observatoire de haute Provence 75
occultation, lunar 83
One-Mile Telescope 85
Orion nebula (M42) 25, 30, 55
Oschin Schmidt Telescope 24, 53, 72
Overwhelmingly Large Telescope (OWL) 124

P

pair production 57–8, 131
parallax 30, 110
Parkes radio telescope 83
Parsons, William 25, 30
Penzias, Arno and Robert Wilson 114
Perlmutter, Saul with Brian Schmidt and Adam Reiss 70
phased array 85, 88
phased-array Feed (PAF) 127
photoelectric effect 11, 39, 52, 57–8
photography 3–4, 21, 50ff.
photometry 51, 54, 112
photometric redshift 73–4, 125
photon optics 12
pixel 47, 53, 62, 64, 73, 125, 133
Planck, Max 11
Planck Space Telescope 118
planetary transit (exoplanet discovery method) 75–6, 109–10
plasma 40, 42–3, 88, 96–7, 101, 112, 116, 118
polarization 41–2, 87, 117–19, 129
prime focus 22–3, 26, 37
proper motion 49, 110
proportional counter 58
pulsar 88ff., 96, 98, 103–4, 119–20

Q

quantum efficiency (QE) 52
quantum nature of light 11, 57, 59, 91, 123
quasars 38, 84, 88, 98, 101, 103, 111, 119

R

radial velocity 34–6, 61, 74–6, 91, 110
Rayleigh scattering 56
Reber, Grote 77–8
redshift 33ff., 61, 66, 70, 73–4, 84, 91–2, 98, 100, 106, 122, 125–6, 128–9
refraction 7, 13ff., 16, 20–1, 24
refractive index 13ff.
robotic telescope 73–4, 76, 125
ROSAT X-ray satellite 101–2
Ryle, Martin 78, 83, 85, 89

S

Sagittarius A* 69
scattering 42, 56, 57–8, 97, 112, 117
scintillation 45, 88
counter/detector 58, 98
Scheiner, Cristophe 28
Schmidt, Bernard 23
Schmidt camera 23–4, 109
Schmidt-Cassegrain Telescope (SCT) 23–4
Schmidt, Maarten 83
Seebeck, Thomas 54
seeing 45
Shapley, Harlow 31–2
shot noise 123
sidereal time 25, 77
Slipher, Vesto 34
Sloan Digital Sky Survey (SDSS) 53, 74, 108, 125, 134
Smith, George 52
Smoot, George 117
software telescope 3, 5, 91ff., 128
Solar and Heliospheric Observatory (SOHO) 111–12
solar flare 78
solar wind 88, 112
Southern African Large Telescope (SALT) 66–8
Space Shuttle 98, 105
spark chamber 58, 98, 100
spectral resolution 59, 73
spectrograph 36, 61–2, 73, 75, 129
spectroscopy 54, 59ff., 74, 91ff., 103, 106, 112, 124
spectrum 4, 12, 21, 34, 41ff., 56, 59ff., 87, 96ff.
speculum 22, 24
Spitzer, Lyman J. Jr. 104–5
Spitzer Space Telescope (SST) 112, 120
Square Kilometre Array (SKA) 93, 126ff., 134
standard candle 30–1, 70
starburst galaxies 94, 112–13
Subaru telescope 26–7, 68–9
sunspots 51, 78, 80, 83
supernova 69–70, 74, 89, 98, 100–1, 103–4, 106, 111, 119, 125, 132
Swift satellite 100
synchrotron radiation 87, 97, 117, 119–20, 129

T

Talbot, William Henry Fox 50
Taylor, Joe 90
thermocouple 54
thermopile 54–5
Thirty Metre Telescope (TMT) 64, 124
transition edge superconducting (TES) bolometers 57, 93–4
two-degree field (2dF) survey 73

U

Uhuru satellite 101
ultra-luminous infrared galaxy (ULIRG) 112
Uranus 24, 106

V

Van der Hulst, Hendrik 91
Vela satellites 97
very large telescope (VLT) 62, 64, 68, 69ff., 72, 74, 86, 106
  interferometer (VLTI) 69
very long baseline interferometry (VLBI) 87ff.
virial mass 36, 101

W

wave optics 12
wavefront 12ff., 45, 60–1, 70–2, 80, 84–5
white dwarf 70, 104
Wide-Angle Search for Exoplanets (WASP) 76
Wide-Field Infrared Survey Telescope (WFIRST) 126
Wien, Wilhelm 42–3
Wilkinson Microwave Anisotropy Probe (WMAP) 117
Wilson, Raymond 67
Wilson, Robert 114
Wolter, Hans 101–2
WorldWide Telescope (WWT) 134–5

X

X-ray Multi-Mirror Mission (XMM-Newton) 102–3

Y

Yerkes refractor 29, 64
Young, Thomas 9–10, 80

Z

Zooniverse 74, 110, 134
Zwicky, Fritz 36, 89

SOCIAL MEDIA
Very Short Introduction

Join our community

www.oup.com/vsi

- Join us online at the official Very Short Introductions **Facebook** page.
- Access the thoughts and musings of our authors with our online **blog**.
- Sign up for our monthly **e-newsletter** to receive information on all new titles publishing that month.
- Browse the full range of Very Short Introductions online.
- Read **extracts** from the Introductions for free.
- If you are a teacher or lecturer you can order inspection copies quickly and simply via our website.

ONLINE CATALOGUE

A Very Short Introduction

Our online catalogue is designed to make it easy to find your ideal Very Short Introduction. View the entire collection by subject area, watch author videos, read sample chapters, and download reading guides.

http://global.oup.com/uk/academic/general/vsi_list/

GALAXIES

A Very Short Introduction

John Gribbin

Galaxies are the building blocks of the Universe: standing like islands in space, each is made up of many hundreds of millions of stars in which the chemical elements are made, around which planets form, and where on at least one of those planets intelligent life has emerged. In this *Very Short Introduction*, renowned science writer John Gribbin describes the extraordinary things that astronomers are learning about galaxies, and explains how this can shed light on the origins and structure of the Universe.

www.oup.com/vsi